MAKING PEACE WITH THE PLANET

ALSO BY BARRY COMMONER

Science and Survival

The Closing Circle

The Poverty of Power

The Politics of Energy

MAKING PEACE WITH THE PLANET

Barry Commoner

THE NEW PRESS
NEW YORK

FOR LISA, WITH LOVE

Portions of this work were originally published in *The New Yorker*, *Harper's* Magazine, *Ramparts*, and *The Columbia Journal of Environmental Law*.

"Failure of the Environmental Effort" by Barry Commoner was originally published in *The Environmental Law Reporter*.

LIBRARY OF CONGRESS CATALOGING-IN-PUBLICATION DATA
Commoner, Barry, 1917–
Making peace with the planet / Barry Commoner.
p. cm.
Includes bibliographical references.
ISBN 1-56584-012-7
1. Pollution—Environmental aspects. 2. Man—Influence on nature.
3. Environmental protection. 4. Nature conservation. I. Title.
QH545.A1C64 1990
304.2'8—dc20

Book Design by Beth Tondreau Design

Manufactured in the United States of America

9 8 7 6 5 4 3 2

CONTENTS

CONTENTS

INTRODUCTION TO THE PAPERBACK EDITION

THE WORLD HAS changed with bewildering speed in the two years since the first edition of this book appeared. The Soviet Union, once one of the world's two superpowers, has been dismembered. Its now separate republics are in the midst of a chaotic economic makeover, elevating the simplistic notion of a planless economy to the sacred status once enjoyed by their rigorously planned one. In the United States, in violation of the Constitution, a war was secretly organized and brutally executed in Iraq, largely in support of our nonexistent energy policy.

The content and temper of public affairs has drastically changed as well. Two years ago, the twentieth anniversary of the first Earth Day firmly established environmental concerns at the top of the public agenda. Many corporations were so convinced that their customers had gone green that they gave up their longstanding posture of denial and favored the once-despised environmentalists with a vulgar embrace—albeit with one eye cocked on the expected profits from their new, supposedly "environmentally friendly" products.

Today, however, environmental issues are nowhere to be seen in the numerous polls that herald an election year, which instead signal intense public concern with unemployment, homelessness, the cost of health care, and the need for a fundamental change in the faulty economy. And George Bush, once the self-anointed "environmental president," now insists that he "cares about the economy."

Why, then, bother to reissue a book about the environment when the country's anguish is centered on a devastating recession that threatens to descend into depression? One reason, of course, is that the environmental crisis is still with us. Another reason is that the government is now less disposed to do anything about it. Following Earth Day 1970, far-reaching remedial actions—new environmental laws and administrative agencies—were initiated. In 1990, despite the acknowledged failure of the country's environmental strategy, Earth Day came and went, leaving scarcely a trace in the country's life apart from admonitions to "keep the lint screen in the clothes dryer clean; be creative with leftovers; go barefoot"—to quote examples gleaned from one of the numerous published "how-to" lists. But there is a far more important reason to think about the environmental issue today: it is a key to understanding the country's *economic* predicament and what can be done about it.

The condition of the environment and the state of the economy are closely connected because both have the same origin: production. *Making Peace with the Planet* and the earlier *The Closing Circle* present an array of evidence that shows how the assault on environmental quality results from the use of systems of production that yield useful goods but also generate pollutants. The engines that power modern cars and trucks also produce pollutants that turn into smog and acid rain. Once disseminated, the synthetic pesticides and chemical fertilizers heavily used in agricultural production also become toxic pollutants that poison our water and food. As the petrochemical industry produces plastics, detergents, and numerous other synthetic chemicals, it also generates almost as much in toxic emissions. Fuel-burning and nuclear power plants produce electricity; but the one also emits gases that are overheating the planet, and the other leaves huge stores of radioactive waste that will outlive the industry itself. The current environmental crisis is the unanticipated outcome of the ways in which the corporations that are entrusted with these decisions have chosen to provide us with transportation, food, and power.

Although most economists like to talk about money—in the form of prices, wages, taxes, and deficits—it is useful to remember that all of these manifestations of wealth, like pollution, are created by production. If the environment is polluted and the economy is sick, the virus that causes both will be found in the system of production. And that is where their cure can be found as well.

As this book demonstrates, there are readily available, environmentally sound production technologies that can replace the environmentally hazardous ones. They include electric cars that do not produce smog-generating nitrogen oxides, or for that matter any other pollutants; photovoltaic cells that generate electricity directly from sunlight, eliminating the pollution emitted by conventional power plants; organic farming that rids the environment of toxic agricultural chemicals; natural materials, such as paper, cotton, or wood, that diminish the need for the synthetic products of the environmentally hazardous petrochemical industry.

The substitution of such ecologically sound production technologies for polluting ones can also improve economic productivity. For example, studies show that organic farmers can obtain the same economic return per acre as conventional farmers, but with a lower debt burden. If

American auto manufacturers stopped whining about their Japanese competitors and instead moved first into mass production of electric cars, they might recapture their once prominent position in the world market. If the United States produced enough reasonably priced photovoltaic cells to satisfy the huge existing market in developing countries, our balance of trade could be improved.

But none of this will happen without government leadership. The electric car is a good example. Not long ago General Motors exhibited a prototype electric car, but announced that it would not be produced until annual sales of 100,000—enough to justify the investment in the necessary factory—could be assured. But since the car is not on sale, there is no demand, creating an economic paralysis that blocks this very desirable environmental and economic breakthrough. Yet, as a recent study by the Center for the Biology of Natural Systems (CBNS) shows, the federal, state, and local governments could break this deadlock with orders for 100,000 electric cars that could be included among the 135,000 passenger vehicles that they buy annually for the fleets they operate.

Another CBNS study shows how Federal purchases could rapidly expand the market for the small, languishing industry that manufactures photovoltaic cells. This simple device can be used to charge the rechargeable batteries that readily replace dry cells. Our calculation showed that a photovoltaic/rechargeable battery system costing about $15 million annually could replace the dry cells that the Federal government purchases at an annual cost of $120 million. Moreover, this purchase would about double the capacity of the U.S. photovoltaic-cell industry, which in turn would reduce the price by a third, opening up even larger markets—for example for roadside lighting. With this one money-saving move, the government could set off a progressive expansion of a new ecologically sound, economically productive industry.

The country's longstanding demand for environmental improvement and the newer calls for economic reform can be realized only by transforming our major systems of production—in industry, agriculture, power production, and transportation. But these decisions are in private, largely corporate hands and until now have been inaccessible to the American people. In the United States, democracy stops at the door to the corporate boardroom. But the door can be opened by the government's massive purchasing power—an economically potent but thus far idle instrument of American democracy. For the sake of the environment, the economy, and of democracy itself, it is time that we used it.

B.C.
January 1992

1

AT WAR WITH THE PLANET

PEOPLE LIVE in two worlds. Like all living things, we inhabit the natural world, created over the Earth's 5-billion-year history by physical, chemical, and biological processes. The other world is our own creation: homes, cars, farms, factories, laboratories, food, clothing, books, paintings, music, poetry. We accept responsibility for events in our own world, but not for what occurs in the natural one. Its storms, droughts, and floods are "acts of God," free of human control and exempt from our responsibility.

Now, on a planetary scale, this division has been breached. With the appearance of a continent-sized hole in the Earth's protective ozone layer and the threat of global warming, even droughts, floods, and heat waves may become unwitting acts of man.

Like the Creation, the portending global events are cosmic: they change the relationship between the planet Earth and its star, the sun. The sun's powerful influence on the Earth is exerted by two forces: gravity and solar radiation. Gravity is a nearly steady force that fixes the planet's path around the sun. Solar radiation—largely visible and ultraviolet light—is a vast stream of energy that bathes the Earth's surface, fluctuating from day to night and season to season. Solar energy fuels the energy-requiring processes of life; it creates the planet's climate and governs the gradual evolution and the current behavior of its huge and varied population of living things. We have been tampering with this powerful force, unaware, like the Sorcerer's Apprentice, of the potentially disastrous consequences of our actions.

We have become accustomed to the now mundane image of the Earth as seen from the first expedition to the moon—a beautiful blue sphere decorated by swirls of fleecy clouds. It is a spectacularly natural object; at that distance, no overt signs of human activity are visible. But this image, now repeatedly thrust before us in photographs, posters, and advertisements, is misleading. Even if the global warming catastrophe never materializes, and the ozone hole remains an esoteric, polar phenomenon, already human activity has profoundly altered global conditions in ways that may not register on the camera. Everywhere in the world, there is now radioactivity that was not there before, the dangerous residue of nuclear explosions and the nuclear power industry; noxious fumes of smog blanket every major city; carcinogenic synthetic pesticides have been detected in mother's milk all over the world; great forests have been cut down, destroying ecological niches and their resident species.

As it reaches the Earth's surface, solar radiation is absorbed and sooner or later converted to heat. The amount of solar radiation that falls on the Earth and of the heat that escapes it depends not only on the daily turning of the Earth and the yearly change of the seasons, but also on the status of the thin gaseous envelope that surrounds the planet. One of the natural constituents of the outer layer of Earth's gaseous skin—the stratosphere—is ozone, a gas made of three oxygen atoms (ordinary oxygen is made of two atoms). Ozone absorbs much of the ultraviolet light radiated from the sun and thereby shields the Earth's surface from its destructive effects. Carbon dioxide and several other atmospheric components act like a valve: they are transparent to visible light but hold back invisible infrared radiation. The light that reaches the Earth's surface during the day is converted to heat that radiates outward in the form of infrared energy. Carbon dioxide, along with several other less prominent gases in the air, governs the Earth's temperature by holding back this outward radiation of heat energy. The greater the carbon dioxide content of the atmosphere, the higher the Earth's temperature. Glass has a similar effect, which causes the winter sun to warm a greenhouse; hence, the "greenhouse effect," the term commonly applied to global warming.

These global effects are not new; they have massively altered the condition of the Earth's surface over its long history. For example, because the early Earth lacked oxygen and therefore the ozone shield, it was once so heavily bathed in solar ultraviolet light as to limit living things to dark places; intense ultraviolet radiation can kill living cells and induce cancer. Similarly, analyses of ice (and the entrapped air bubbles) deposited in the Antarctic over the last 150,000 years indicate that the Earth's temperature fluctuated considerably, closely paralleled by changes in the carbon dioxide level.

Changes in the Earth's vegetation can be expected to influence the carbon dioxide content of the atmosphere. Thus, the massive growth of forests some 200 million to 300 million years ago took carbon dioxide out of the air, eventually converting its carbon into the deposits of coal, oil, and natural gas produced by geological transformation of the dying trees and plants. The huge deposits of fossil fuel, the product of millions of years of photosynthesis, remained untouched until coal, and later petroleum and natural gas, were mined and burned, releasing carbon dioxide into the atmosphere. The amounts of these fuels burned to provide human society with energy represent the carbon captured by photosynthesis over millions of years. So, by burning them, in the last 750 years we have returned carbon dioxide to the atmosphere thousands of times faster than the rate at which it was removed by the early tropical forests. The atmosphere's carbon dioxide content has increased by 20 percent since 1850, and there is good evidence that the Earth's average temperature has increased about 1 degree Fahrenheit since then. If nothing is done to change this trend, temperatures may rise by about 2.5 to 10 degrees more in the next fifty years. This is about the same change in temperature that marked the end of the last ice age about 15,000 years ago—an event that drastically altered the global habitat. If the new, man-made warming occurs, there will be equally drastic changes, this time endangering a good deal of the world that people have fashioned for themselves. Polar ice will melt and the warmer oceans will expand, raising the sea level and flooding many cities; productive agricultural areas, such as the U.S. Midwest, may become deserts; the weather is likely to become more violent.

Regardless of how serious the resultant warming of the Earth turns out to be, and what, if anything, can be done to avoid its cataclysmic effects, it demonstrates a basic fact: that in the short span of its history, human society has exerted an effect on its planetary habitat that matches the size and impact of the natural processes that until now solely governed the global condition.

The ozone effect leads to the same conclusion. This problem arises not from the rapid man-made reversal of a natural process, but from the intrusion of an unnatural one on global chemistry. The chief culprits are the synthetic chemicals known as CFCs or chlorofluorocarbons. Like most of the petrochemical industry's products, CFCs do not occur in nature; they are synthesized for use in air conditioners, refrigerators, and spray cans, as solvents, and as a means of producing foam plastics. CFCs readily evaporate and are extraordinarily stable; escaping from confinement in a junked air conditioner or a discarded plastic cup, they migrate upward into the stratosphere. There they encounter ozone molecules, generated by the impact of solar radiation on ordinary oxygen molecules. A complex catalytic reaction ensues, in which each CFC molecule causes the destruction of numerous ozone molecules. This chemical process has already eaten a huge hole in the protective ozone layer over Antarctica, evidence that here, too, a process recently created by human society matches in scope a natural, protective component of the Earth's global envelope. Serious damage to people, wildlife, and crops is likely if the process continues: a large increase in skin cancer; eye problems; suppression of photosynthesis. Moreover, the CFCs act like carbon dioxide toward heat radiation and, along with methane and several minor gases, contribute to global warming.

Clearly, we need to understand the interaction between our two worlds: the natural ecosphere, the thin global skin of air, water, and soil and the plants and animals that live in it, and the man-made technosphere—powerful enough to deserve so grandiose a term. The technosphere has become sufficiently large and intense to alter the natural processes that govern the ecosphere. And in turn, the altered ecosphere threatens to flood our great cities, dry up our bountiful farms, contaminate our food and water, and poison our bodies—catastrophically diminishing our ability to provide for basic human needs. The human attack on the ecosphere has instigated an ecological counterattack. The two worlds are at war.

The two spheres in which we live are governed by very different laws. One of the basic laws of the ecosphere can be summed up as "Everything is connected to everything else." This expresses the fact that the ecosphere is an elaborate network, in which each component part is linked to many others. Thus, in an aquatic ecosystem a fish is not only a fish, the parent of other fish. It is also the producer of organic waste that nourishes microorganisms and ultimately aquatic plants; the consumer of oxygen produced photosynthetically by the plants; the habitat of parasites; the fish hawk's prey. The fish is not only, existentially, a fish, but also an element of this network, which defines its functions. Indeed, in the evolutionary sense, a good part of the network—the microorganisms and plants, for example—preceded the fish, which could establish itself only because it fitted properly into the preexisting system.

In the technosphere, the component parts—the thousands of different man-made objects—have a very different relation to their surroundings. A car, for example, imposes itself on the neighborhood rather than being defined by it; the same car is sold for use on the densely packed Los Angeles freeways or in a quiet country village. It is produced solely as a salable object—a commodity—with little regard for how well it fits into either sphere: the system of transportation or the environment. It is true, of course, that all cars must have a width that is accommodated by the traffic lanes, and must have proper brakes, lights, and horn, and so on. But as every resident of Los Angeles or New York knows, in recent years their crowded streets and highways have been afflicted with longer and longer limousines, designed to please the buyer and profit the producer, but hardly suitable to their habitat.

Defined so narrowly, it is no surprise that cars have properties that are hostile to their environment. After World War II, the American car was arbitrarily redefined as a larger, heavier object than its predecessors. That narrow decision dictated a more powerful engine; in turn, this required a higher engine compression ratio; in keeping with physical laws, the new engines ran hotter; at the elevated temperature, oxygen and nitrogen molecules in the cylinder air reacted chemically, producing nitrogen oxides; leaving the engine exhaust pipe, nitrogen oxides trigger the formation of the noxious smog that now envelops every major city. The new cars were successfully designed to carry people more comfortably at higher speed; but no attention was paid to an essential component in their habitat—the people themselves, and their requirement for clean, smog-free air.

Even a part of the technosphere as close to nature as the farm suffers from the same sort of clash with the environment. As a man-made object, the farm is designed for the sole purpose of producing crops. Guided by that purpose, after World War II agronomists urged the increasingly heavy application of chemical nitrogen fertilizer. Yields rose, but not in proportion to the rate of fertilizer application; year by year, less and less of the applied fertilizer was taken up by the crop and progressively more drained through the soil into groundwater, in the form of nitrate that contaminated rivers, lakes, and water supplies. Nitrogen fertilizer is a commodity sold with the narrow purpose of raising yields and manufactured with the even narrower purpose of increasing the chemical industry's profits. When inorganic nitrogen fertilizer was introduced in the 1950s, little or no attention was paid to its ecological behavior in the soil/water system or to the harmful effects of elevated nitrate levels in drinking water.

The second law of ecology—"Everything has to go somewhere"—together with the first, expresses the fundamental importance of cycles in the ecosphere. In the aquatic ecosystem, for example, the participating chemical elements move through closed cyclical processes. As they respire, fish produce carbon dioxide, which in turn is absorbed by aquatic plants and is used, photosynthetically, to produce oxygen—which the fish respire. The fish excrete nitrogen-containing organic compounds in their waste; when the waste is metabolized by aquatic bacteria and molds, the organic nitrogen is converted to nitrate; this, in turn, is an essential nutrient for the aquatic algae; these, ingested by the fish, contribute to their organic waste, and the cycle is complete. The same sort of cycle operates in the soil: plants grow, nourished by carbon dioxide from the air and nitrate from the soil; eaten by animals, the crop sustains their metabolism; the animals excrete carbon dioxide to the air and organic compounds to the soil—where microorganisms convert them into compounds such as nitrate, which nourish the crop. In such a closed, circular system, there is no such thing as "waste"; everything that is produced in one part of the cycle "goes somewhere" and is used in a later step.

The technosphere, in contrast, is dominated by *linear* processes. Crops and the animals to which they are fed are eaten by people; their waste is flushed into the sewer system, altered in composition but not in amount at a treatment plant, and the residue is dumped into rivers or the ocean as waste—which upsets the natural aquatic ecosystem. Uranium is mined, processed into nuclear fuel which, in generating power, becomes highly radioactive waste that must be carefully guarded—ineffectually thus far—from contaminating the environment for thousands of years. The petrochemical industry converts ethylene prepared from petroleum and chlorine prepared from brine into vinyl chloride, a synthetic, carcinogenic chemical. This is manufactured into the plastic, polyvinyl chloride, which is made into tile, boots, and food wrapping; sooner or later discarded, these become trash that must be disposed of. When burned in an incinerator, the polyvinyl chloride produces carbon dioxide and dioxin; both are injected, as waste, into the ecosphere where the one contributes to global warming and the other to the risk of cancer. The energy sources that now power the technosphere are mostly fossil fuels, stores that, once depleted, will never be renewed. The end result of this linear process is air pollution and the threat of global warming. Thus, in the technosphere goods are converted, linearly, into waste: crops into sewage; uranium into radioactive residues; petroleum and chlorine into dioxin; fossil fuels into carbon dioxide. In the technosphere, the end of the line is always waste, an assault on the cyclical processes that sustain the ecosphere.

The third informal law of ecology is "Nature knows best." The ecosystem is consistent with itself; its numerous components are compatible with each other and with the whole. Such a harmonious structure is the outcome of a very long period of trial and error—the 5 billion years of biological evolution. The biological sector of the ecosphere—the biosphere—is composed of living things that have survived this test because of their finely tuned adaptation to the particular ecological niche that they occupy. Left to their own devices, ecosystems are conservative; the rate of evolution is very slow, and temporary changes, such as an overpopulation of rabbits, for example, are quickly readjusted by the wolves.

The same sort of conservative self-consistency governs the chemical processes that occur in living cells. For example, there are severe constraints imposed on the variety of organic (carbon-containing) compounds that are the basic components of biochemical processes. As the physicist Walter Elsasser has pointed out, the weight of one molecule of each of the proteins that *could* be formed from the twenty different amino acids that comprise them would be greater than the weight of the known universe. Obviously, living things are constrained to produce only a very small number of the *possible* proteins. Constraints are also exercised by the enzymes, present in all living things, that catalyze the degradation of organic compounds. It is an unbroken rule that for every organic compound produced by a living thing, there is somewhere in the ecosystem an enzyme capable of breaking it down. Organic compounds incapable of enzymatic degradation are not produced by living things. This arrangement is essential to the harmony of the ecosystem. If, for example, there were no enzymes that degrade cellulose, an otherwise very stable major constituent of plant cell walls, the Earth's surface would eventually become buried in it.

Similarly, certain molecular arrangements are shunned in the chemistry of life. Thus, very few chlorinated organic compounds, in which chlorine atoms are attached to carbons, occur in living things. This suggests that the vast number of chlorinated organic compounds that are possible chemically (many of them now produced by the petrochemical industry), have been rejected in the long course of evolution as *biochemical* components. The absence of a particular substance from nature is often a sign that it is incompatible with the chemistry of life. For example, the fact that mercury plays no biochemical role and does not normally occur in living cells—and is lethal when it does—is readily explained by the fact that it poisons a number of essential enzymes. In the same way, many man-made chlorinated organic compounds that do not occur in nature, such as DDT or dioxin, are very toxic.

In sum, the living things that comprise the biosphere, and their chemical composition, reflect constraints that severely limit their range of variation. The mermaid and the centaur are, after all, mythical animals; even the vaunted exploits of genetic engineering will never produce an elephant-sized mouse or a flying giraffe. In the same way, no natural biochemical system includes DDT, PCB, or dioxin. Unfortunately, these highly toxic substances are not mythical—a fact that sharply illuminates the difference between the ecosphere and the technosphere.

In contrast to the ecosphere, the technosphere is composed of objects and materials that reflect a rapid and relentless process of change and variation. In less than a century, transport has progressed from the horse-drawn carriage, through the Model T Ford, to the present array of annually modified cars and aircraft. In a not much longer period, writing instruments have evolved from the quill pen to the typewriter and now the word processor. Synthetic organic chemistry began innocuously enough about 150 years ago with the laboratory production of a common natural substance—urea—but soon departed from this imitative approach to produce a huge array of organic compounds never found in nature and, for that reason, often incompatible with the chemistry of life. Nylon, for example, unlike a natural polymer such as cellulose, is not biodegradable—that is, there is no enzyme in any known living organism that can break it down. As a result, when it is discarded into the ecosphere, nylon, like plastics generally, persists. Thus, oceanographers now find in their collecting nets bits of orange, blue, and white nylon and larger pieces jammed in the digestive tracts of dead turtles—the residue of nylon marine cordage. In the technosphere, nylon is a useful new commodity; in the ecosphere, nylon, untested by evolution, is a harmful intruder.

"Nature knows best" is shorthand for the view that during the several billion years in which they have evolved, living things have created a limited but self-consistent array of substances and reactions that are essential to life. The petrochemical industry has departed from these restrictions, producing thousands of new man-made substances. Since they are based on the same fundamental patterns of carbon chemistry as the natural compounds, the new ones are often readily accepted into biochemical processes. They therefore can play an insidious, destructive role in living things. For example, synthetic organic compounds may easily fit into the same reactive enzyme niches as natural molecules or may be accepted into the structure of DNA. However, they are sufficiently different from the natural compounds to then disrupt normal biochemistry, leading to mutations, cancer, and in many different ways to death. In effect, the petrochemical industry produces substances that—like the fantasies of human society invaded by look-alike but dangerous aliens—cunningly enter the chemistry of life, and attack it.

Finally, it is useful to compare the ecosphere and the technosphere with respect to the consequences of failure. In the ecosphere, this is expressed by the idea that "there is no such thing as a free lunch," meaning that any distortion of an ecological cycle, or the intrusion of an incompatible component (such as a toxic chemical), leads unavoidably to harmful effects. At first glance, the technosphere appears to be extraordinarily free of mistakes—that is, a technological process or product that failed not because of some unanticipated accident but because it was unable to do what it was designed to do. Yet nearly every modern technology has grave faults, which appear not as a failure to accomplish its designed purpose but as a serious impact on the environment. Cars usually run very well, but produce smog; power plants efficiently generate electricity, but also emit dangerous pollutants; modern chemical farming is very productive but contaminates groundwater with nitrate and wildlife and people with pesticides. Even the spectacular nuclear disasters at Three Mile Island and Chernobyl were far less serious as technical failures than they were in their ecological effects. Regarded only as a failure in the plant's function, the accident at Chernobyl amounts to a serious but local fire that destroyed the plant. But the resultant release of radioactivity threatens many thousands of people all over Europe with cancer. In sum, there are numerous failures in the modern technosphere; but their effects are visited upon the ecosphere.

A free lunch is really a debt. In the technosphere, a debt is an acknowledged but unmet cost—the mortgage on a factory building, for example. Such a debt is tolerable because the technosphere is a system of production, which—if it functions properly—generates goods that represent wealth potentially capable of repaying the debt. In the technosphere, debts are repaid from within and, at least in theory, are always capable of being paid off, or, in some cases, canceled. In contrast, when the debts represented by environmental pollution are created by the technosphere and transferred to the ecosphere, they are never canceled; damage is unavoidable. The debts represented by the radioactivity disseminated from the nuclear accident at Chernobyl, and by the toxic chemicals that enveloped Bhopal, have not been canceled. These debts were merely transferred to the victims, and are paid as they sicken and die.

Since they inhabit both worlds, people are caught in the clash between the ecosphere and the technosphere. What we call the "environmental crisis"—the array of critical unsolved problems ranging from local toxic dumps to the disruption of global climate—is a product of the drastic mismatch between the cyclical, conservative, and self-consistent processes of the ecosphere and the linear, innovative, but ecologically disharmonious processes of the technosphere.

Since the environmental crisis has been generated by the war between the two worlds that human society occupies, it can be properly understood only in terms of their interplay. Of course, as in a conventional war, the issues can be simplified by taking sides: ignoring the interests of one combatant or the other. But this is done only at the cost of understanding. If the ecosphere is ignored, it is possible to define the environmental crisis solely in terms of the factors that govern the technosphere: production, prices, and profits, and the economic processes that mediate their interaction. Then, for example, one can concoct a scheme, as recently proposed by President Bush, in which factories are allotted the right to emit pollutants up to some acceptable level and, in a parody of the "free market," to buy and sell these rights. But unlike the conventional marketplace, which deals in goods—things that serve a useful purpose—this scheme creates a marketplace in "bads"—things that are not only useless but often deadly. Apart from the issue of morality, it should be noted that such a scheme cannot operate unless the right to produce pollutants is exercised—hardly an inducement to eliminating them.

If the technosphere is ignored, the environmental crisis can be defined in purely ecological terms. Human beings are then seen as a peculiar species, unique among living things, that is doomed to destroy its own habitat. Thus simplified, the issue attracts simplistic solutions: reduce the number of people; limit their share of nature's resources; protect all other species from the human marauder by endowing them with "rights."

This approach raises a profound, unavoidable moral question: Is the ecosphere to be protected from destruction for its own sake, or to enhance the welfare of the human beings who depend on it? This leads to a further question regarding the term "welfare." Some environmental advocates believe that human welfare would be improved if people were less dependent on the artifacts of the technosphere and lived in closer harmony with their regional ecosystem—baking bread instead of buying it; walking or pedaling a bike instead of driving a car; living in small towns instead of cities. The thrust of this approach is to deny the value to society of, let us say, a woman who uses time saved by buying bread instead of baking it in order to work as a curator in an urban museum. Nor does it allow for the possibility that time- and labor-saving technologies can be compatible with the integrity of the environment. It assumes that the technosphere, no matter how designed, is necessarily an environmentally unacceptable means of giving people access to resources that are not part of their ecological niche. But as we shall see, this assumption is wrong; although nearly every aspect of the *current* technosphere is counterecological, technologies exist that—although little used thus far—*are* compatible with the ecosphere.

The view that people are to be regarded *solely* as components of the ecosystem can lead to extreme and often inhumane proposals. Consider the global warming issue, for example. The humanist approach dictates a vigorous effort to halt the process because it is a massive threat to human society: flooded cities, drought-ridden agriculture, and prolonged heat waves. However, judged only in ecological terms, global warming can be regarded merely as a change in the structure of the global ecosystem similar to the warming that accompanied the last postglacial period, albeit more rapid. Viewed in this way, there is no more reason to oppose global warming than to be unhappy about the last ice age and the rise in global temperature that ended it. At its farthest reach, this nonhumanist position becomes antihumanist, as exemplified in an article in the publication of a group called Earth First!, which favored the spread of AIDS as a means of reducing the human population without threatening other animal species. Of course, at the other extreme is the potentially suicidal view that the enormous value of modern production technology to human society justifies whatever damage to the ecosphere it entails.

The ambiguity created by the dual habitat in which we live has led to a very wide range of responses. The extreme interpretations of the relationship between the two spheres that human society occupies—and a sometimes bewildering array of intermediate positions—is compelling evidence that we have not yet understood how the two systems have come into conflict and, as a result, are unable as yet to resolve that conflict.

This book is an effort to analyze the war between the ecosphere and the technosphere, written with the conviction that understanding it—as distinct from reacting to it—is the only path to peace. It is less a lament over the war's numerous casualties than an inquiry into how future casualties can be prevented. It is not so much a battle cry for one side or the other, as a design for negotiating an end to this suicidal war—for making peace with the planet.

2

THE ENVIRONMENTAL FAILURE

IF, AS WE HAVE SEEN, the environmental crisis is generated by the clash between the ecosphere and the technosphere, it becomes imperative that we learn how they interact and what can be done to harmonize them. A useful way to analyze such a relationship is to alter one of the interacting systems and then observe what changes occur in the other. Such a test has, in fact, been carried out in the United States and a number of other countries on a grand scale for the last twenty years. Beginning in 1970, numerous measures were adopted to improve the quality of the environment. Major components of the technosphere—automobiles, power plants, and petrochemical factories—were changed; they were required to adopt control devices in order to reduce their impact on the ecosphere. A good deal can be learned, therefore, by examining the ecological improvement generated by these technological alterations. Such a review has the secondary benefit of evaluating the effectiveness of the movement that has urged this campaign. The modern environmental movement is old enough now—its birth can also be dated from the enthusiastic outburst of Earth Day in April 1970—to be accountable for its successes and failures. Having made a serious claim on public attention and the nation's resources, the movement's supporters and the responsible government agencies cannot now evade the troublesome, potentially embarrassing question: What has been accomplished?

The United States is a good place to look for answers. Concern with the environment and efforts to improve it are now worldwide, but the United States is the place where the environmental movement first took hold, and where the earliest efforts were made. Since the early 1970s, the country has been governed by basic laws that were intended to eliminate air and water pollution and to rid the environment of toxic chemicals and of agricultural and urban wastes. National and state environmental agencies have been established; about a trillion dollars of public and private money have been spent; powerful environmental lobbies have been created; local organizations have proliferated. Environmental issues have taken a permanent place in the country's political life.

In one respect, all this activity has clearly achieved an important success: we now know much more about the state of the environment than we used to. Since the early 1970s, the United States has established monitoring and reporting systems that record the annual emissions of pollutants and changes in environmental quality. These give us a picture of what has happened in the environment since the effort to improve it began.

There are several useful ways of evaluating the environment that differ in their ease of measurement and in their relevance to the measure of ultimate interest: the effect on people and other living things. The amounts of different pollutants that are emitted into the environment each year can be estimated fairly accurately for the nation as a whole from technical data on the behavior of cars, power plants, factories, or farms. But such measurements fail to reveal the levels of pollutants that people actually encounter, which may differ a great deal depending on their distance from the pollutant's source. Local concentrations of some common pollutants—for example, the amount of dust or sulfur dioxide per cubic meter of air—are measured by a network of air-sampling devices that are stationed at fixed places, chiefly in cities. But the resultant information is spotty and not readily converted to an average national trend. Water, taken from rivers, lakes, underground sources, or wells, is also analyzed from time to time for chemical and biological pollutants, but again the results are necessarily discrete and localized. Finally, there are less complete measurements that determine how much of certain pollutants, such as pesticides, have been taken up by wildlife, especially fish and birds; there are also a few corresponding analyses of pollutants carried in the bodies of the human population.

The nation's basic environmental law, the National Environmental Policy Act of 1969, assigns the task of reporting these data to the Council on Environmental Quality (CEQ). Unfortunately, one of the Reagan administration's first of many cuts in domestic programs was a sharp reduction in the CEQ budget (its staff has been reduced to ten), which has resulted in diminished reports, especially on water pollution. Nevertheless, by rounding out the CEQ reports with special ones produced by various government agencies, it is possible to piece together a picture of how the environment has fared in the last fifteen to twenty years.

Information about the trends in air pollution is available from annual reports published by the Environmental Protection Agency (EPA) since 1975. (Data earlier than 1975 tend to be unreliable because measurements were not standardized.) The reports describe changes in the emissions and local concentrations of the major airborne pollutants: particulates (dust), sulfur dioxide, lead, nitrogen oxides, volatile organic compounds, and ozone, a key ingredient of photochemical smog. One striking fact is immediately evident from the data: it is indeed possible to reduce the level of pollution sharply, for between 1975 and 1987 total annual lead emissions decreased by 94 percent and airborne concentrations at national test sites by 92 percent. Lead is a notoriously toxic metal that causes serious health effects, such as mental retardation, especially among children living in heavily polluted areas. Children have benefited from the reduced emissions. The average lead levels in children's blood decreased by 37 percent between 1976 and 1980.

The successful effort to reduce lead pollution only accentuates the failure to achieve a comparable reduction in the emissions of all the other air pollutants. On the average, the emissions of particulates, sulfur dioxide, carbon monoxide, nitrogen oxides, and volatile organic compounds decreased by only 18 percent between 1975 and 1987. Of these pollutants, dust emissions have improved most—about 33 percent—although actual concentrations in the air at some 1,510 sites have improved somewhat less. But the annual improvements came to a halt in 1982.

Sulfur dioxide is a particularly serious air pollutant, for it diminishes the respiratory system's ability to deal with all other pollutants. It is also a major contributor to acid rain. Between 1975 and 1987, total sulfur dioxide emissions declined by 20 percent. Average concentrations at national test sites improved somewhat more, in part because new power plants—a major source of sulfur dioxide—are being built outside urban areas, where most of the test sites are located.

Carbon monoxide is chiefly produced by cars, trucks, and buses, and the effort to control it is based on a device—the catalytic converter—that destroys carbon monoxide before it is emitted with the automobile exhaust. (The control device also reduces waste fuel emissions.) Total annual carbon monoxide emissions decreased by 24 percent between 1975 and 1987.

Photochemical smog is a complex mixture, created when nitrogen oxides emitted from automobile exhausts and power plants are converted by sunlight into highly reactive molecules that then combine with waste fuel and other hydrocarbons to form ozone and other noxious chemicals. The total emissions of nitrogen oxides *increased* by 2 percent between 1975 and 1987. Ozone is not emitted as such but is formed in the air during the smog reactions.

The noxious smog chemicals are responsible for serious health hazards; people with heart or respiratory problems are routinely warned to stay indoors during "smog alerts." This hazard is now more or less accepted as an apparently unavoidable aspect of urban life. In some places, improvements in smog levels have been achieved by reducing traffic. Yet smog continues to threaten health. For example, in Los Angeles, the worst-afflicted city, between 1973 and 1977 residents were subjected each year to an average of 250 days on which smog was at levels classified as "unhealthful"; 150 days were classified as "very unhealthful." In most U.S. cities, residents are still exposed to unhealthful smog levels for 50 to 150 days each year. On sunny, windless days the telltale brown haze of smog can be seen hanging in the air over nearly every American city. Smog is a continuing hazard to the health of urban residents.

One of the consequences of the unsolved problems of air pollution is acid rain. In keeping with the ecological law "Everything has to go somewhere," once emitted into the air, sulfur dioxide and nitrogen oxides are picked up by rain and snow and brought down to earth in the form of sulfate and nitrate. Both of these substances increase acidity, and in recent years, many lakes, especially in the northeastern

United States and Canada, as well as in Europe, have become more acid. In some of these lakes, there have been serious biological changes, often involving the virtual elimination of fish populations. Forest growth has been reduced by acid rain. In cities acid rain erodes buildings and monuments.

Scrubbers installed at coal-burning power plants have somewhat reduced the plants' sulfur dioxide emissions, and since emissions of nitrogen oxides from power plants and automotive vehicles have increased, there has been a noticeable shift in the relative contributions of these pollutants to acid rain. Reports from Hubbard Brook, a research station in New Hampshire, where acidity problems have been studied for a long time, show that while the sulfate content of precipitation fell by 25 percent between 1964 and 1981, the nitrate content increased by 137 percent. Not much improvement in the acid rain problem can be expected, given the negligible improvement in sulfur dioxide emissions and the rising emissions of nitrogen oxides, especially from automobiles.

Thus, since 1975 the overall improvement of most air pollutant emissions has been at best modest. The only exception is lead; emissions have decreased year by year, from 147,000 metric tons in 1975 to only 8,100 in 1987. Moreover, the annual reduction in the emissions of four of the standard air pollutants—particulates, sulfur dioxide, nitrogen oxide, and volatile organic compounds—came to a halt in 1982. Since then, there has been no statistically significant change in annual emissions, suggesting that the effort to reduce them has reached its limit.

The ecological processes that govern the quality of surface waters are more complex than the chiefly chemical events that govern air pollution. A basic reason for the pollution of surface waters—rivers and lakes—as well as inshore marine waters is the stress placed on the natural ecological cycles which, if they are kept in balance, maintain water quality. If inadequate sewage treatment systems dump excessive organic matter into a river or a lake, the accompanying fecal bacteria threaten health. As the excess organic matter is broken down by aquatic microorganisms, the organisms may consume so much oxygen that fish begin to die. Urban and industrial waste and runoff from agricultural areas—and indeed the normal effluent from sewage treatment plants—increase nitrate and phosphate concentrations. If they rise beyond the levels maintained by a balanced aquatic cycle, eutrophication occurs. Heavy algal blooms are formed and soon die, burdening the system with excessive organic matter and reducing oxygen content. In addition, high nitrate levels in drinking water may create health problems such as methemoglobinemia (a condition that reduces the oxygen-carrying capacity of the blood, especially in infants) and may contribute to the formation of carcinogens. Toxic chemicals add to these harmful effects.

In the last decade, particular rivers and lakes here and there have been cleaned up to a degree by closing sources of pollution and building new sewage treatment plants. Yet in that period, nationally, there has been little or no overall improvement in the levels of the five standard pollutants that determine water quality: fecal coliform bacteria, dissolved oxygen, nitrate, phosphate, and suspended sediments. A recent U.S. Geological Survey report on the trends in pollution levels between 1974 and 1981 at nearly four hundred locations on major American rivers shows that there has been no improvement in water quality at more than four-fifths of the tested sites. For example, the levels of fecal coliform bacteria decreased at only 15 percent of the river stations, and increased at 5 percent. At half the locations, the bacterial count was too high to permit swimming, according to the standard recommended by the National Technical Advisory Committee on Water Quality Criteria. Levels of dissolved oxygen, suspended sediments, and phosphorus improved at 13 to 17 percent of the locations, but deteriorated at 11 to 16 percent of them. The most striking change—for the worse—was in nitrate levels: increases were observed at 30 percent of the test stations and decreases at only 7 percent. Agricultural use of nitrogen fertilizer is a main source of this pollutant; in rivers that drain cropland, the number of sampling stations that report rising nitrate levels is eight times the number reporting falling levels. Another major source is nitrogen oxides emitted into the air by vehicles and power plants and deposited in rain and snow as nitrate; this accounts for increased river nitrate levels in the Northeast, despite the relative scarcity of heavily fertilized acreage in that area. The survey also shows that there was a sharp increase in the occurrence of two toxic elements, arsenic and cadmium (a cause of lung and kidney damage), in American rivers between 1974 and 1981; but, as expected from the reduced automotive emissions, the occurrence of lead declined.

An overall assessment of the changes in these standard measures of water quality can be gained from the average trends. For the five standard pollutants, the frequency of improving trends averaged 13.2 percent; but the frequency of deteriorating trends averaged 14.7 percent; thus, at more than four-fifths of the test sites, overall water quality deteriorated or remained the same. In sum, the regulations mandated by the Clean Water Act, and more than $100 billion spent to meet them, have failed to improve water quality in most rivers. A few places have improved, but more have deteriorated. Moreover, the presence of at least three serious pollutants—nitrate, arsenic, and cadmium—has increased considerably. Nor is there any evidence in these data that pollution levels will improve; like the effort to clean up the air, the campaign to reduce river pollution has stalled, after reaching only a very modest level of success.

One of the chief symptoms of the environmental crisis in the early 1970s was eutrophication, especially in lakes. Lake

Erie was a dramatic example. The lake received an enormous burden of inadequately treated sewage and phosphate-rich detergents from the cities that surround it, and chemical plants contributed mercury and other toxic materials as well. Rivers carried nitrate and eroded soil from heavily fertilized farms into the lake. By the 1960s and 1970s, especially in its western regions and along the shoreline, Lake Erie was exhibiting the classic signs of eutrophication: heavy algal overgrowths, epidemics of asphyxiated fish, and sharply declining fish catches.

Because of its notoriety as a "dying lake," serious efforts have been made to revive Lake Erie. It has also been intensively studied in the last decade, and the results evaluated by elaborate statistical techniques. The most detailed data are described in a recent lengthy EPA report. The report makes a telling comparison of pollution levels along various reaches of the western lakeshore, where eutrophication has been particularly troublesome. In 1972–73, three of twenty-one shoreline regions were classified as entirely or partially eutrophic; in 1978–79 eutrophication was more widespread, affecting twelve of twenty-one shoreline regions.

The rate at which oxygen is removed from lake water by decay processes is an important indicator of water quality. Rapid oxygen depletion is evidence of heavy pollution with organic matter, and is, of course, a threat to the oxygen-requiring fish. According to the EPA report, the rate of oxygen depletion in the central basin of Lake Erie increased by 15 percent between 1970 and 1980, following a trend that goes back as far as 1930. Apparently, the processes that have been overburdening the lake with organic matter since 1930 have continued. The entry of phosphorus from certain city sewage systems has declined somewhat as a result of campaigns to reduce the use of phosphate-containing detergents. Probably the most successful effort, in Detroit, reduced river phosphate concentrations by 70 percent between 1971 and 1981. But the acquisition of phosphate from all of the rivers that enter the lake has been reduced much less; only four of twelve test sites showed any improvement. Overall, phosphate entering Lake Erie decreased by about a third between 1972 and 1982. The recent EPA report also records the catch of commercial fish. The once valuable catch of herring, whitefish, pike, and walleye dropped off sharply after 1950, and by 1960 was close to zero. It failed to recover between 1960 and 1980.

In 1970 Lake Erie was widely mourned as a dying lake; now, nearly two decades after our environmental reawakening, despite some limited improvements, it remains a flagrant example of environmental pollution. The condition of less famous lakes is just as bad. This is particularly true of lakes in heavily farmed areas, where nitrogen fertilizer leaches from the soil into rivers and lakes. A 1982 survey found that of 107 lakes in Iowa, all were eutrophic—and so were more than 80 percent of the lakes in Ohio and Pennsylvania.

About 50 percent of the population of the United States depends on underground sources of water—groundwater—for its drinking water. The U.S. Geological Survey and state agencies monitor the quality of groundwater by testing wells throughout the country. The results based on testing more than 100,000 wells show that in the past twenty-five years these sources are becoming increasingly polluted by nitrate and toxic chemicals. Fertilizer is chiefly responsible for the rising nitrate levels. In Nebraska a 1983 survey showed that 82 percent of the wells over the nitrate limit established by health authorities (10 milligrams per liter of nitrogen in the form of nitrate) were contaminated by fertilizer nitrogen. In California's Sacramento Valley, a very heavily farmed area, nitrate contamination of wells has been followed for a long time. In the fifty-year period following 1912, the percentage of wells with excessive nitrate (defined as 5.5 milligrams of nitrogen per liter) approximately doubled. More recently, the percentage of wells with excessive nitrate doubled again—this time in only a four-year period, between 1974 and 1978. The major source of the nitrate is nitrogen fertilizer leaching from irrigation water. A similar trend has been observed in Iowa.

In 1984 the U.S. Geological Survey summarized the situation: "Current trends suggest that nitrate accumulations in ground water of the United States will continue to increase in the future." Clearly, we have failed to solve this environmental problem, which grows worse with time.

Fifteen years ago, public opinion polls on environmental issues showed that most people were worried about air pollution—especially smog. Now, even though these problems remain largely unsolved, polls show that as a public concern, air and water pollution run well behind a new environmental threat—toxic chemicals. For the first time in the 4-billion-year history of life on the planet, living things are burdened with a host of alien man-made substances that are harmful to them. In the early 1970s, this problem was largely due to agricultural products—insecticides, herbicides, and fungicides. DDT and similar chlorinated insecticides were the most notorious examples. In 1972 the use of DDT and related insecticides was banned in the United States (although production and exports to developing countries continue) because they were shown to promote cancer and also to be a hazard to wildlife. One of the most noticeable effects of the use of DDT was the decline in bird populations; DDT interferes with the biochemistry of reproduction, making eggs thin-shelled and thus easily destroyed before they hatch. Banning the use of DDT has been very effective. For example, between 1969 and 1975, the average DDT content of brown pelicans in South Carolina decreased by 77 percent and by 1976 the number of fledglings more than tripled. People have benefited as well: between 1970 and 1983 average DDT levels in body fat in the U.S. population decreased by 79 percent. The banning of polychlorinated biphenyl

(PCB), another notorious chemical pollutant—it increases the incidence of cancer and birth defects—has had a similar effect. Between 1970 and 1980, the body burden of PCB in fresh-water fish decreased by 56 percent, and in starlings, by 86 percent. In people, the percentage of the population with relatively high levels of PCB (above 3 parts per million) in their fatty tissue decreased by about 75 percent.

Unfortunately, the improved environmental levels of DDT and PCB are the exception. Since 1950 the roster of serious chemical pollutants has steadily expanded. Hundreds of toxic chemicals, many of them carcinogenic, have persisted in water supplies, air, and food. According to a recent EPA survey, people in the general U.S. population now carry dozens of man-made synthetic chemicals, many of them carcinogenic, in their body fat and in the fat of mothers' milk as well.

A particularly unfortunate feature of chemical pollution is that it is often created unwittingly. Perhaps the most striking example is the group of related compounds known popularly as "dioxin"—more accurately, a family of 75 polychlorinated paradibenzodioxins and 135 polychlorinated dibenzofurans. The chief dioxin hazard appears to be its extraordinary ability to enhance the incidence of cancer. In animal experiments, it increases the incidence of cancer at dose levels appreciably lower than any other synthetic compound. The present environmental levels of dioxin—as revealed by the amounts recently detected by the EPA in the adipose tissue (fat) of nine hundred people representative of the general U.S. population—create a maximum lifetime cancer risk of about 330 per million. This is well over the EPA guideline of 1 per million and greater than the maximum lifetime cancer risk expected from the population's exposure to benzene, which has been subjected to regulation under the Clean Air Act on that account.

Even more serious is the effect of dioxin exposure on breast-fed infants. The dioxin content, per unit fat, in breast milk is about the same as that found in body fat. On that basis, an infant breast-fed for only one year already has a lifetime chance of getting cancer of about 48 in a million. The major source of environmental dioxin is not the various chemicals (such as the herbicide 2, 4, 5-T) in which it occurs as a contaminant, but a seemingly innocent process—the incineration of urban trash (and sewage sludge as well). It turns out that dioxin is actually synthesized in such a trash-burning incinerator by a reaction between a common constituent of paper and wood (and therefore of trash)—lignin—and chlorine derived from the combustion of chlorinated plastic, such as polyvinyl chloride. This is something new; trash incineration has been going on for many years, but according to analyses of dated sediments in the Great Lakes, dioxin first appeared in the U.S. environment between 1930 and 1940, increasing a great deal since then. The increase parallels the production of chlorinated organic chemicals such as polyvinyl chloride by the petrochemical industry. About one-fourth of the polyvinyl chloride is used for packaging; discarded into the trash stream, it has turned the incinerators into dioxin factories.

The saga of dioxin is dramatic, but it is typical of the new, unexpected hazards created by the massive production of man-made chemicals. There are many other examples. For a long time, chlorine has been used in water purification plants to kill dangerous bacteria. Now the purification process has itself become dangerous, because chlorine reacts with the numerous chemicals that now contaminate our water supplies, producing chloroform and other carcinogens.

The total toxic chemical problem is huge. According to the recent EPA Toxic Release Inventory, U.S. industry emits about 20 billion pounds of toxic chemicals annually into the environment. However, according to Congress's Office of Technology Assessment, because of underreporting and the omission of data from small establishments, this figure is very uncertain and is more likely to be about 400 billion pounds. Only about 1 percent of the chemical industry's toxic waste is actually destroyed. The chemical industry, its emissions of toxic substances largely unrestrained, has become the major threat to environmental quality.

Unfortunately, unlike standard air and water pollutants, there are no periodic estimates made of the levels of toxic pollutants in the environment; indirect evidence indicates that the trend is upward. For example, birth defects in Great Lakes birds, which are due to toxic chemicals, increased thirty-one–fold between the early 1970s and 1978.

Recently, new concerns have been raised about the impact of chemical contamination on the safety of the nation's food supply, especially for children. Since the toxic chemicals emitted into the environment occur in air, drinking water, and food, they readily enter the human body. Most of them are especially soluble in fat and therefore tend to accumulate in fatty components of the body. The EPA adipose tissue survey tested for the presence of thirty-seven toxic compounds—all of them synthetic products of the petrochemical industry. All but four of the compounds were found, generally in about two-thirds or more of the fat samples tested. Prominent among the detected substances are carcinogens such as benzene and chloroform. As noted earlier, the compounds found in body fat can be expected to occur in breast milk as well. According to a recent report of the National Research Council in 1987, the pesticide residues in food could account for as many as 20,000 additional cases of cancer per year in the U.S. population.

Environmental pollution from radioactive materials is in a class by itself; it originates from a single sector of production—the manipulation of nuclear energy for peaceful or military purposes. Since World War II, the United States has created a large industrial establishment to produce nuclear weapons. Only recently, the government has been forced to acknowledge that many of these operations have created serious radioactive hazards in the surrounding areas. The peaceful uses of nuclear energy are encompassed by nuclear power production and the use of radioactive materials in research, medical diagnosis, and in certain industrial operations. The amounts of radioactive material involved in the nuclear power industry are much greater than the amounts involved in the other uses. Radiation exposure due to the *normal* operation of the nuclear power industry is quite small when compared with exposure to uncontrollable, natural sources of radioactivity such as cosmic rays; it is less than 0.01 percent of the natural exposure. But this is a national average; near power plants, exposures—and the resultant medical risks—are certainly higher.

Nuclear power plants and the associated fuel-handling operations may release substantial amounts of radioactive material into the environment in abnormal circumstances: a traffic accident that spills radioactive waste on a road; a stuck or improperly handled valve at a nuclear power plant that vents radioactive gas into the air or radioactive water into a river; leakage of radioactive waste from storage tanks. There is no widely disseminated accounting of the radioactivity that enters the environment in these ways. At most, there are only reports that an accidental emission of radioactive material has occurred. Generally, there is no information about the actual radiation exposure, but only the customary statement "the amount of radiation released was harmless." (Such statements are wrong. Radiation, no matter how weak, always involves the risk of some harm, and the risk is proportional to the dose received. A very small exposure from a dental X-ray has a correspondingly small risk, which is presumably worth taking in view of the expected benefit.) A partial measure of the radioactivity released by nuclear power plants is available from EPA studies of krypton 85, a radioactive gas uniquely associated with the operation of such plants. The average annual concentration increased by 80 percent between 1970 and 1983, thus increasing the environmental hazard.

Accidents that disrupt the containments protecting the environment from the huge amount of radioactive material in an operating nuclear reactor are a far larger hazard than normal operations. Until 1979 this was an abstract issue, argued by appeals to abstruse statistical computations which concluded that the probability of a serious accident might be as low as one in 100 million. The accident at the Three Mile Island nuclear power plant in 1979 brought this discussion down to earth. While the outcome of the accident was far less serious than it might have been—a partial fuel meltdown extensively contaminated the interior of the reactor building, and released some radioactivity into the environment—it suggested that a serious mechanical failure was much more probable than the calculations had indicated.

In April of 1986, of course, a second failure occurred—at the Soviet Union's Chernobyl nuclear power plant—which led to a radioactive disaster. More than 250 people have died from acute radiation sickness; estimates of future cancer deaths from the radioactive fallout over a wide area of Europe range well above 100,000; more than 100,000 people were evacuated from their homes, and recently more evacuations have been planned; some hundreds of square miles of agricultural land have become useless; radioactive fallout disrupted milk and vegetable production over most of Europe.

Both the accidents at Three Mile Island and Chernobyl appear to have had a similar origin: some failure in the normal heat-dispersing system that led to sudden overheating and a runaway reaction. In both accidents hydrogen was produced, which exploded at Chernobyl, but fortunately at Three Mile Island did not. Like Three Mile Island, the Chernobyl reactor was encased in a containment structure, but it failed in the explosion. U.S. containment structures are somewhat different, but are also subject to failure. It is worth noting that the two reactors were under very different systems of management. In the Soviet Union, design, construction, and operation of nuclear power plants is highly centralized and uniform from one facility to another. In the United States, reactors are made to order by several different manufacturers, and they are operated by different power companies. That these basically similar accidents have occurred under such diverse systems of management suggests that they reflect a fault that is inherent in the technology itself, and which is probably incapable of being cured by managerial skill. The fault is not some esoteric, highly improbable event. (The head of the Atomic Energy Administration once equated it to the probability of being bitten by a snake while crossing a street in Washington, D.C.) Rather, it is an event that, on the basis of the actual record, may happen, with potentially disastrous consequences, once every seven years or so.

With the accidents at Three Mile Island and Chernobyl, nuclear power—a presumably "peaceful" technology, spawned by the deplorable, infinitely dangerous technology of nuclear war—has reached its own unhappy maturity. During the effort to improve the environment over the last decade, nuclear power has manifested its proclivity for malfunctions that threaten enormous damage to the environment—a threat that has been realized at Chernobyl. Like many other environmental hazards, in the period since the birth of the movement that hoped to remove them, the environmental impact of nuclear power has become more dangerous.

The hazardous consequence of one aspect of nuclear technology—fallout from test explosions of nuclear weapons in the atmosphere—has been considerably reduced by the straightforward procedure of simply stopping the process that creates it. As a result of the 1963 treaty between the United States and the Soviet Union, atmospheric tests have halted (except for a few conducted by China and France) and fallout radioactivity has declined, in keeping with natural decay processes, about tenfold. For example, a national survey of milk showed that it contained 23.8 picocuries of strontium 90 per liter in 1964, 7.3 in 1970, and 2.3 in 1983.

Although European environmental data are generally less complete than the U.S. data, they follow the same pattern. Thus, in West Germany the average reduction in the emission of the standard air pollutants declined by only 15.4 percent between 1970 and 1982—a figure quite close to the average U.S. decline. And, as in the United States, the nitrogen oxide picture was worst—*increasing* by 29 percent. In Great Britain, the average decline in emissions of the standard air pollutants was less than 1 percent; emissions of carbon monoxide and volatile organic compounds increased by more than 10 percent. Scattered data from the socialist bloc suggest that air quality has deteriorated considerably in the last twenty years: Poland, for example, experienced a 358 percent increase in nitrogen oxide emissions between 1978 and 1982.

A recent report from the Global Environment Monitoring System, operated by the United Nations Environment Programme and the World Health Organization since 1974, confirms that the failure to improve urban air quality is worldwide. Thus, between 1973 and 1985, in nineteen developed countries, the average improvement in dust emissions was only 7.5 percent; in fourteen such countries the average improvement in sulfur dioxide emissions was only 12.8 percent. In several of these countries, notably Greece, Israel, Poland, and Portugal, such emissions *increased* considerably. According to the report, between 1973 and 1983 nitrogen oxide emissions in Belgium, France, Japan, and Greece were either unchanged or increased slightly. Between 1974 and 1983, in most countries carbon monoxide emissions improved about as much as they did in the United

States; however, in Ireland, Israel, and Hong Kong, there was no improvement. As in the United States, those countries that have reduced the lead content of gasoline report significantly lower lead emissions, notably Sweden and Ireland.

There is also a parallelism between U.S. and European data on water pollution. For example, environmental changes in the Baltic Sea closely resemble those in Lake Erie; because of continued phosphate and nitrate pollution, eutrophication persists and oxygen levels are low. In the Baltic Sea, between 1979 and 1984 average oxygen levels decreased by 11 percent, while phosphate concentrations increased by 101 percent and nitrate concentrations increased by 37 percent. And, as in Lake Erie, for apparently the same reasons, the levels of DDT and PCB in fish improved considerably, by 80 percent and 45 percent respectively.

The failure to prevent the continued deterioration of the environment—let alone improve it—is, of course, painfully evident on the global scale. All that can be said about global warming is that in the last few years its potentially catastrophic consequences have at least been widely recognized. Thus far, nothing has been done to prevent or minimize this potential disaster. Ozone depletion has also continued to worsen. The Reagan administration's initial response, as voiced by his secretary of energy, Donald Hodel, was to promote the use of dark glasses and suntan lotion. Since then, some more positive steps have been taken: an international agreement to reduce CFC production by 50 percent has been reached; in some places—Irvine, California, for example—local reductions in CFC emissions have been enacted into law; the Du Pont chemical company has announced the development of a less damaging substitute for CFCs; in August 1989, several automobile companies promised to reduce the losses of CFCs from automobile air conditioners. But none of these measures are enough; the threat of a disastrous increase in ultraviolet radiation is nearly undiminished.

The most recent issue of the Council on Environmental Quality's report, *Environmental Quality, 1987–1988*, summarizes these trends in levels of pollution. It confirms the evidence cited earlier and adds to it the following: Between 1971 and 1985, at eighteen areas of the U.S. coast, the proportion of shellfish waters unavailable for use because of pollution increased from 45 percent to 49 percent. Between 1970 and 1985 the number of oil spills in and around U.S. waters increased by 196 percent and their total volume by 57 percent. Between 1974 and 1984 the number of hazardous waste spills increased thirty-four-fold and their total volume by 239 percent.

In addition to these harmful intrusions into the fragile web of the ecosphere, natural resources have been destroyed at an increasing rate. Perhaps the most serious problem is the destruction of virgin forests, especially in the tropics. Such forests support a huge diversity of animal and plant species; when the forests are destroyed on a large scale, their resident species die and may be lost forever. In Latin America, for example, it has been estimated that at the present rate of clearing, nearly half the original forest area will be lost by the end of the century, and with it about 15 percent of the original plant and animal species.

How can we judge these results of the twenty-year effort to clean up the environment? What standards are applicable? At least in the area of air pollution, specific goals have been established by EPA's interpretation of the enabling legislation. But clearly, except for lead, we have failed to even approximate the 90 percent improvements mandated in 1970 by the Clean Air Act. In 1971 EPA published a detailed rule-making document that establishes national air quality standards, defines the automotive emissions controls needed to achieve them, and specifies the emission reductions to be accomplished. The predicted emission levels took into account the efficiency of the prescribed exhaust control devices, their expected deterioration with use, the projected numbers of cars and trucks, and the number of miles they would travel annually. According to these predictions, between 1975 and 1985 annual carbon monoxide emissions were expected to decline by 80 percent and nitrogen oxide emissions by 70 percent. They were far off the mark. In reality, in that period carbon monoxide emissions fell by only 19.1 percent, and nitrogen dioxide emissions *increased* by 4.2 percent. By this clear-cut standard, the effort to reduce the serious environmental impact of automotive vehicles must be judged a failure.

In the face of such failures, enforcement has marched backward, in a scandalous retreat. In 1970 the Clean Air Act Amendments set a 1977 deadline for achieving a 90 percent reduction in urban carbon monoxide, hydrocarbon, and ozone levels. The penalty for failure is severe: loss of federal funding for development projects. In 1977, with compliance not even in sight, the deadline was moved to 1982; and when that was also missed by many municipalities, the deadline was once more delayed, to December 31, 1987—and was missed by a number of cities, in which nearly 150 million people still breathe substandard air. Since then, the three most polluted cities—Los Angeles, New York, and Houston—have been given twenty more years to comply, a delay of thirty-two years from the original deadline.

The summer of 1988 was a kind of memorial to the failure of the nation's environmental program. Smog reached unprecedented levels; on the East Coast, numerous beaches were closed by an influx of sewage debris and medical waste; the emergence of the pollution of Boston Harbor as an election issue was added evidence that a problem as old and fundamental as sewage disposal remained unsolved; everywhere in the country, communities struggled with the trash disposal problem, as landfills were closed and costs mounted;

and the record-breaking heat was an alarming reminder that nothing has been done to combat the potential greenhouse effect.

In sum, the Congress has mandated massive environmental improvement; the EPA has devised elaborate, detailed means of achieving this goal; most of the prescribed measures have been carried out, at least in part; and in nearly every case, the effort has failed to even approximate the goals. In both the columns of statistics and everyday experience, there is inescapable evidence that the massive national effort to restore the quality of the environment has failed.

3

PREVENTION VERSUS CONTROL

HOW CAN WE ACCOUNT for the failure of the massive effort to clean up the environment—that apart from a very few exceptions, pollutant emissions have decreased only modestly, if at all? One can argue, of course, that the original expectation of a 90 percent improvement was unrealistic and that the modest rate of decline in standard air pollutant emissions (apart from lead) of about 1 percent per year—a rate that has dropped to zero since 1982—is about the best we can do. But that argument is contradicted by the fact that the sought-for improvement *has* been achieved for a handful of pollutants: airborne lead, DDT and related pesticides, PCB, mercury in surface waters, radioactive fallout from nuclear bomb tests, and in some rivers, phosphate. The sharp divergence between these few successes and the many failures creates an opportunity for understanding both. We can ask: Which remedial measures have led to failure, and which to success?

The impact of a pollutant on the environment can be remedied in two general ways: either the activity that generates the pollutant is changed to eliminate it; or, without altering the activity, a control device is added that traps or destroys the pollutant before it can enter the environment. Thus, without otherwise changing the automobile, a catalytic converter has been attached to its exhaust to destroy carbon monoxide and unused gasoline, and conventional power plants have been equipped with scrubbers that trap sulfur dioxide and dust. As we have seen, such appended controls have failed to achieve the mandated improvements. Some "controls"—for example, the common practice of pumping hazardous chemical wastes into deep, water-bearing strata—are only a kind of ecological sleight of hand, for they merely secrete the pollutant for a time in a less noticeable part of the environment. Sooner or later, they, too, lead to failure.

The few real improvements have been achieved not by adding controls or concealing pollutants but by simply eliminating them. The reason there is so much less lead in the environment—and in children's blood—is that lead has been almost entirely eliminated from the manufacture of gasoline. The reason why DDT and related pesticides are now much less prevalent in wildlife and our own bodies is that their use has been banned. The levels of mercury in rivers and lakes have greatly declined because it is no longer used in chlorine production. Phosphate concentrations have been sharply reduced in some rivers because local legislation has banned its use in detergents. And there is now much less strontium 90 in milk and in children's bones than there was twenty years ago, because we and the Russians have had the simple wisdom to halt the atmospheric testing of nuclear bombs, which produces it.

In sum, there is a common explanation for each of the few sharp reductions in emissions: in each case, environmental degradation was *prevented* by simply stopping production or use of the pollutant. This suggests an addition to the informal environmental laws: If you don't put something into the environment, it isn't there.

Controls yield little or no improvement in environmental quality because they are ultimately self-defeating. Viewed only as a separate device, an environmental control may appear to be highly efficient: in order to meet current air pollution standards, a catalytic converter is designed to trap 96 percent of the exhaust's carbon monoxide; and a power plant scrubber can trap 70 to 90 percent of the plant's sulfur dioxide emissions. Surely these devices should be capable of substantially reducing emissions. But in practice, the devices are only one element in a larger system that can readily counteract their apparent efficiency.

To begin with, no control device is ever perfect. Unlike the absolute impact of banning a pollutant—which, after all, reduces emissions to zero—the installation of a control system cannot completely halt pollution. For example, the catalytic converter's effectiveness rapidly declines with use. Tests of catalytic converters in operating cars show that up to 25,000 miles of use, 45 percent of the cars meet the carbon monoxide emission standard; between 25,000 and 50,000 miles of use, 28 percent meet the standard; and with more than 50,000 miles of use, only 10 percent meet the standard. Most important, because the control device is not perfect, continued increase in the pollution-generating activity (traffic, for example) will gradually overwhelm the device's limited ability to improve environmental quality. Finally, control devices are completely useless unless the pollutant enters the environment at only one or a few points, where they can be installed. The classic example of such a "nonpoint source" is the nitrate that leaches from heavily fertilized soil into groundwater from every square inch of soil. There is no conceivable way of controlling this process; as long as nitrogen fertilizer is used at levels that exceed the rate at which it is absorbed by the crop, nitrate will continue to leach into groundwater and the rivers that drain agricultural areas.

In sum, a control device always allows some pollution to enter the environment, so that increased productive activity negates the device's intended effect. In contrast, when a pollutant is simply eliminated or banned, its rate of entry into the environment falls permanently to zero. If pollution is prevented, environmental quality is compatible with increased economic activity; if pollution is controlled, they clash. In the task of restoring environmental quality, prevention works; control does not.

Prevention succeeds because it is directed at the *origin* of the pollutant in the production process itself—the vast and varied machinery of industry, agriculture, transportation, and manufacturing. We can then see that each of the few successful environmental improvements has been achieved by altering the technology of production. The very considerable reduction in mercury in Lake Erie nicely illustrates this precept. In 1970 there were very high concentrations of this toxic metal in Lake Erie sediments; most of the lake bottom contained 1,000 to 2,000 parts per million of mercury and more than that near the outlet of the Detroit River. In March 1970, pickerel contaminated with 7 parts per million of mercury—fourteen times above the acceptable level—were found in a tributary of the lake. The major source was soon discovered: chlorine-producing plants that used mercury to conduct the electric current which, passed through a brine solution, yields the chlorine. Threatened by legal action, the plants adopted an alternate technology in which a semipermeable diaphragm instead of mercury establishes the electric circuit. By 1979 most of the lake sediments contained less than 300 parts per million of mercury and fish levels were much improved. Mercury pollution in Lake Erie was sharply curtailed simply by changing the technology of chlorine production to eliminate the use of mercury.

The same approach is responsible for the sharp reduction in phosphate emissions from urban sewage systems. In the early 1970s, eutrophication—the proliferation of algal overgrowths that quickly die and pollute the water—received a good deal of public attention. In many places, it was found that the process was triggered by excess levels of phosphate—an algal nutrient—which had been added to synthetic detergents as a dirt-suspending agent. After 1970, when some cities enacted regulations requiring a sharp reduction in the phosphate content of detergents, the manufacturers substituted other agents and river phosphate levels declined. Thus, phosphate pollution was sharply reduced, at least locally, by changing the technology of detergent production.

PCB provides a similar example. This synthetic, highly toxic substance was widely used in a number of processes ranging from the manufacture of "carbonless" carbon paper to electric transformers. When PCB was banned in 1979, each of these processes was altered to exclude it, and emissions into the environment fell precipitously. In the same way, the levels of DDT and related insecticides in the environment and in the human body have declined a great deal because they were banned and eliminated from agricultural practice in the early 1970s. DDT had been used chiefly in cotton production; when it was banned, the technology of cotton production was changed. In this case, one hazard was replaced by another. Toxaphene, also a hazardous insecticide, was introduced to replace DDT; as a result, while environmental levels of DDT declined, the toxaphene content of fish increased more than twentyfold between 1970 and 1980. The technology of cotton production is reflected in the quality of the environment. Finally, environmental strontium 90 has decreased because the technology of nuclear bomb testing was changed when the 1963 test ban treaty eliminated testing in the atmosphere.

Thus, there is a consistent explanation for the few instances of environmental success: they occur only when the relevant production technologies are changed to eliminate the pollutant. If they are not changed, pollution continues unabated, or at best—if a control device is added—it is only slightly reduced. In effect, the effort to deal with environmental pollution has been trivialized. A great deal of attention has been paid to designing control devices—and enforcing their use—that can only moderately reduce hazardous emissions, and are eventually overwhelmed by growth in production. Much less attention has been given to the more difficult but far more rewarding task of changing the basic technologies that produce the pollutants. We now know that environmental pollution is an incurable disease; it can only be prevented. But instead of preventing the underlying disease, Band-Aids have been applied.

That the few instances of environmental improvement result from changes in the technology of production confirms earlier evidence (which I have reviewed in *The Closing Circle*) that the same systems of production that have generated the nation's enormous wealth are also responsible for the present, excessive levels of environmental pollution. Most of our current production technologies are only forty to fifty years old, the result of a massive transformation in agriculture, transportation, power production, and manufacturing that began after World War II. With very few exceptions—chiefly television, computers, and other electronic marvels—this technological transformation has not produced new *kinds* of goods. What has changed is the *way* in which goods are produced.

Before 1950 crops were grown without chemical nitrogen fertilizer or synthetic pesticides; now these chemicals have become a major element in crop production. Before 1950

American cars were small and driven by low-compression engines; now they are larger, heavier, with higher engine compression ratios. Before 1950 beer and soda were sold in reusable bottles; now they are sold in containers that are used once and then converted into trash. Before 1950 cleansers were made of soap; now over 85 percent are synthetic detergents. Before 1950 clothes were made of natural fibers—cotton, wool, silk, and linen; now man-made, synthetic fibers have captured a large share of the market. Before 1950 all these goods were shipped from farm and factory to distant cities by rail; now highway trucks have taken over most freight hauling. Before 1950 meat was wrapped in paper and taken home in a paper bag; now it is encased in plastic and carried home in a plastic bag. Before 1950 college cafeterias and fast-food restaurants used washable plates and utensils; now everything is "disposable," which means that used once, it becomes trash. Before 1950 every baby's bottom was diapered in reusable cotton; now most babies sport throwaway diapers. Before 1950 no one in their right mind would throw out a razor or a camera after using it once; now this is commonplace.

Each of these changes created a new assault on the environment or greatly intensified an old one. What has happened to the American car is perhaps the sorriest example of this process. As some of us remember, the pre–World War II car was a pretty serviceable vehicle; it couldn't make jackrabbit starts or accelerate around curves, but it did the job of carrying us around the country. Its internal combustion engine polluted the air with carbon monoxide and unburned gasoline, hazards that were fairly localized. Then, as the U.S. auto industry closed down its wartime operations and returned to the business of providing Americans with personal cars, a change was made. The new postwar cars were heavier, and therefore necessarily driven by more powerful engines; between 1950 and 1968 the horsepower of the average U.S. car engine increased from 100 to 250. To accomplish this, the engines were redesigned to run at higher cylinder compression ratios, which increased by 50 percent. Obedient to the laws of physics, the higher pressure caused increased engine temperatures. Now the laws of chemistry, which dictate more rapid reactions at higher temperatures, took hold; at the elevated engine temperature, the oxygen and nitrogen in the cylinder air reacted to form nitrogen oxides. Once emitted from the exhaust and exposed to sunlight, nitrogen oxides reacted with hydrocarbons, such as waste gasoline in the air, forming ozone and the noxious mixture known as photochemical smog. Finally, the leftover nitrogen oxide was converted to acidic nitrate and carried to the ground in rain and snow, becoming a major contributor to acid rain.

Because the engine was changed, the gasoline had to be modified as well, creating an added environmental hazard. High-compression engines tend to burn fuel unevenly, causing "engine knock" that greatly reduces the engine's power and durability. Knocking can be prevented by adding tetraethyl lead to the gasoline. As compression ratios increased, it was necessary to add progressively more lead to the gasoline; between 1950 and 1968 the average lead content of U.S. gasoline increased by 35 percent. By 1970 automobiles and trucks were responsible for 80 percent of the amount of this highly toxic metal emitted into the air.

Simply stated, the postwar decision by U.S. auto companies to build larger, more powerful vehicles *created* new environmental hazards: the smog that now envelops every American city in deadly vapors; lead levels in the blood of inner-city children sufficient to account for symptoms of mental retardation; a major contribution to the acid rain that has depopulated numerous lakes of their fish and threatens widespread destruction of forests. A seemingly innocent alteration in the technology of automobile production has spawned a pervasive assault on the environment.

Not only have the vehicles been changed, but also the

transportation system as a whole—again to the detriment of the environment. Truck freight has progressively replaced rail freight. Since a truck burns four times more fuel to move a ton-mile of freight than a railroad—and produces that much more pollution—freight hauling has become more hazardous to the environment and to our health. The substitution of automobiles and buses for intercity and commuter railroads has a similar effect. In all these ways, the post–World War II changes in transportation have intensified environmental impact.

The post–World War II transformation of the American farm tells us a similar story. Between 1950 and 1970, the total U.S. crop output increased by 38 percent, although the acreage decreased by 4 percent and the labor required fell by 58 percent. This sharp increase in productivity was accomplished by an 18 percent increase in the use of machinery and a 295 percent increase in the application of synthetic pesticides and fertilizer.

This technological transformation accounts for major features of the environmental crisis. The sharp increase in the application of nitrogen fertilizer is the chief cause of the progressive rise in the levels of nitrate contamination in rivers and groundwater. Chemical nitrogen fertilizer was introduced to replace the nitrogen available naturally from soil organic matter that had been heavily depleted in many areas, especially the Corn Belt, by poor farming practices. Unlike the natural source of nitrogen in the soil, which is released in soluble form only gradually as the crop grows, the chemical form, which is highly soluble, is applied quickly in large amounts, so that enough will remain in the soil to sustain crop growth. Inevitably, more fertilizer is applied than can be absorbed by the crop, and—in keeping with the law "Everything must go somewhere"—the rest leaches out of the soil into groundwater, rivers, and lakes. Moreover, some of the fertilizer may be converted to gaseous nitrous oxide, which contributes to the greenhouse effect.

The postwar introduction of synthetic pesticides began with DDT and similar chlorinated compounds. Their environmental impact, brought to public attention in Rachel Carson's classic, *Silent Spring,* is notorious: massive fish kills, decimated bird populations, and the risk of cancer in people. Since then, some seven hundred different synthetic chemicals have been introduced into U.S. agriculture to control insects, weeds, and fungi. What all these chemicals have in common is that they kill, and do so at extremely low concentrations. Because they are organic compounds, structurally similar to natural ones, pesticides readily enter into the cell's biochemical systems. However, because they often contain molecular configurations (such as carbon-chlorine bonds) that are rare in natural biochemicals, they disrupt these systems at crucial points. The hazardous effect is not restricted to the targeted insect or weed, for many biochemical processes are common to a very wide range of organisms, where they may play different roles. Thus, the same DDT molecule that kills insects disrupts sex hormone metabolism in birds.

In a fundamental sense, pesticides are similar to pharmaceutical drugs: their intended effect is often accompanied by dangerous side effects. To deal with this problem, drugs are applied person by person in precisely the amounts dictated by studies of their full range of effects. In contrast, huge amounts of pesticides are spread, broadcast, into the environment although little is known about their total impact on it. Very little pesticide actually reaches the intended targets; most unnecessarily exposes wildlife and people to its hazards. Moreover, only 289 of the 700 chemicals currently used in U.S. pesticides have been sufficiently tested to evaluate their side effects; many pesticides that have passed the tests are later found to be harmful. Yet each year 750 million pounds of pesticides are sprayed across the U.S. landscape. The effects on plants, animals, and people can be appreciated by visualizing the consequences of spraying that many pounds of a potpourri of pharmaceutical drugs across the country.

The outcome of this massive uncontrolled experiment in chemical therapy has recently been assessed by the National Research Council of the National Academy of Sciences: 20,000 additional cases of cancer in the U.S. population annually because "the average consumer is exposed to pesticide residues . . . in nearly every food." Nor is it surprising that a study by the National Cancer Institute should find that Kansas farmers who were exposed to herbicides experienced a six-to-eight-fold increase in cancer incidence.

Thus, the major change in the technology of agricultural production since World War II—the massive introduction of chemical fertilizer (especially nitrogen) and synthetic pesticides—is the reason why pollutants have contaminated water supplies; poisoned birds, fish, and wildlife; burdened food with toxic chemicals; and increased the incidence of cancer.

In the petrochemical industry, which produces the toxic chemicals that so grievously pollute the environment, there is a unique relationship between environmental impact and the technology of production. In other sectors of production—for example, agriculture—the product, food, is not itself a pollutant and there are environmentally benign ways of producing it, for example without chemical fertilizers and pesticides. In contrast, toxic wastes are an inescapable accompaniment to the manufacture of almost every petrochemical product. In manufacturing its annual 500 to 600 billion pounds of products, the chemical industry is also responsible for emitting into the environment nearly the same amount of toxic chemicals. The resultant pollution is essentially uncontrolled, for only about 1 percent of the industry's toxic wastes are actually destroyed. The rest enter the environment either immediately or—if they are consigned to underground strata or surface deposits—at some later time. Moreover, many of the industry's products, such as pesticides and plastics, are themselves pollutants as well.

The chemical disaster in Bhopal, India, in 1985, in which more than 2,000 people died and some 200,000 were disabled, many of them for life, dramatizes the problems inherent in chemical production technology. The conventional view is that the Bhopal disaster was due to some technical fault in the plant that produced the toxic chemical, methyl isocyanate, which is used to manufacture insecticides. There is, of course, no perfectly accident-free industrial process; this is particularly true of petrochemical plants, which are not only complex but produce numerous dangerous products and by-products. Indeed, there have been repeated accidents at the Union Carbide methyl isocyanate plants in Bhopal and in Institute, West Virginia, that have exposed workers and in some cases the neighborhood to this dangerous substance.

The conventional view accepts the inevitability of such risks and seeks to balance them against the benefits of producing methyl isocyanate: the jobs created for chemical workers and the food protected by insecticides. But the facts contradict this position. Consider how well the Bhopal plant provided jobs. Bhopal is a city of 700,000 people, with perhaps a third of them living in shanties next to the plant. But only a few hundred worked in the plant. The reason why so many people lived close to the plant is that they came there for jobs in *constructing* it, remaining after the plant was built, and now, out of work, living in extreme poverty.

How well do the pesticides that are produced from methyl isocyanate—carbamate insecticides such as Sevin and Temik—provide more food? Of course, such pesticides can protect crops from insect attack and improve production (although as we shall see in chapter 5, there are better ways of doing this). However, most of the pesticides used in developing countries are not used on local food crops, but on crops grown for export—cotton, coffee, or bananas. Such "monoculture" agriculture is particularly susceptible to insect attack, leading to very heavy reliance on pesticides, especially as the insects become resistant. For example, El Salvador, a very small country, uses about one-fifth of the world's parathion output to treat its coffee crop. Moreover, Third World countries use many pesticides that have been banned in the developed countries because of their biological hazards. According to one survey, India has used nine of the twelve pesticides that have been banned in the United States. Although relatively little pesticide is used directly on local food crops, those crops are nevertheless exposed to pesticides sprayed on nearby export crops. As a result, food grown in Third World countries is often heavily contaminated.

Unlike the steel, auto, or electric power industries, the petrochemical industry—at least on its present scale—is not essential. Nearly all of the products of the petrochemical industry are substitutes for perfectly serviceable preexisting ones. Plastics substitute for paper, wood, and metals; detergents for soap; nitrogen fertilizer for soil organic matter and nitrogen-fixing crops (the natural sources of nitrogen); pesticides for the insects' natural predators. There are, of course, certain petrochemical products that are new, unique, and useful—for example, pharmaceutical drugs, videotape, or the artificial plastic heart—but these represent only a small part of the industry's total output, most of which is made up of textiles, containers, packaging, and other substitute commodities.

As petrochemical substitutes have invaded the economy, they have made everyday activities more hazardous to the environment. Washing was once a rather benign intrusion on the environment; soap, which is made of a natural organic compound, fat, is readily degraded in sewage treatment plants. Now that most cleansers are synthetic detergents, washing has become environmentally hazardous because detergents can cause algal overgrowths and pollute the water with noxious chemical residues. When shopping bags were made of paper, they could be disposed of by composting or recycling, or even simply disintegrated by the weather. Today, discarded plastic shopping bags are either burned—and pollute the air—or are with us forever, most in landfills but, increasingly, festooned in trees. When soda bottles were made of glass, they could be refilled and used dozens of times; now in plastic, the bottles are used once and become trash.

In sum, the petrochemical industry is unique. Not only are its wastes dangerous, but its very products degrade the environment much more than the ones they displace. The petrochemical industry is inherently inimical to environmental quality.

Environmental degradation is built into the technical design of the modern instruments of production. The environmental hazard is just as much an outcome of the facility's technological design as is its productive benefit. High compression is the cause of both the auto engine's power and its production of nitrogen oxide—which triggers smog. The extensive use of nitrogen fertilizer accounts for the high productivity of the modern farm—and for the pollution of rivers and groundwater as well. The same biochemical potency that made DDT an effective insecticide is also responsible for massive fish kills.

In sum, there have been sweeping changes in the technology of production since World War II. Natural products—soap, cotton, wool, wood, paper, and leather—have been displaced by synthetic petrochemical products: detergents, synthetic fibers, and plastics. In agriculture, natural fertilizers—manure and nitrogen-fixing crops—have been displaced by chemical fertilizers; and natural methods of pest control—crop rotation, ladybugs, and birds—have been displaced by synthetic pesticides. In transportation, small cars have been replaced by much larger ones, trolley cars by buses, and rail freight by truck freight. In commerce, reusable goods have been replaced by throwaways. These changes have turned the nation's farms, factories, vehicles, and shops into seedbeds of pollution.

As I pointed out in *The Closing Circle,* by 1970 it was clear that these changes in the technology of production are the *root cause* of modern environmental pollution. Now this conclusion has been confirmed by the sharply divergent results of the effort to clean up the environment. Only in the few instances in which the technology of production has been changed—by eliminating lead from gasoline, mercury from chlorine production, DDT from agriculture, PCB from the electrical industry, and atmospheric nuclear explosions from the military enterprise—has the environment been substantially improved. When a pollutant is attacked at the point of origin—in the production process that generates it—the pollutant can be eliminated; once it is produced, it is too late. This is the simple but powerful lesson of the two decades of intense but largely futile effort to improve the quality of the environment.

4

THE COST OF FAILURE

THE ENVIRONMENTAL EFFORT—a massive but misdirected attempt to solve a major social problem—has failed. A major cost of this failure is, of course, the still polluted environment. And perversely, the national campaign has created new problems instead of solving old ones—a second generation of crippling mistakes.

The U.S. environmental program itself is not only ineffectual but also enormously—and unnecessarily—complex and costly. Preventing a disease, after all, is far more efficient than treating it. Consider, for example, how effectively smallpox has been dealt with as compared with cancer. The battle against smallpox, however costly, is over; with the virus eliminated worldwide from the human population, there will be no further costs. In contrast, the effort to treat cancer after it appears is endless and progressively more costly. Here, too, only prevention really works. The most effective way to combat cancer due to cigarette smoking is not to smoke; that eliminates one sure source of the disease and its subsequent costs. In contrast, if the cause of the disease—smoking—is disregarded, a continuing burden of illness, with its personal anguish, elaborate medical effort, and high public costs, is imposed on society.

In the same way, the failure to address the origins of our environmental problems—the only way to prevent them—has left us with the costly, continuing battle against the symptoms, $365 billion for only the five-year period from 1983 to 1987. The size and complexity of this futile battle is also evident from the administrative machinery that was set in motion by the environmental laws. The Clean Air Act (amended) of 1970 is a good example. The act specifies some forty separate steps that EPA and the fifty states must take to establish air quality standards and procedures for enforcing them. EPA must determine the effects of each air pollutant on public health and welfare, for example by developing evidence that sulfur dioxide damages the lungs' protective mechanism, that it has been implicated in bronchitis and lung cancer, and is partly responsible for the damage of acid rain to forests, crops, and buildings. Thus identified as a "pollutant which has an adverse effect on public health or welfare," sulfur dioxide's biological, chemical, and physical effects are then eligible for detailed study, with the aim of estimating the damage from different levels of exposure.

The EPA can then establish a "primary standard"—levels of sulfur dioxide in air that, for the sake of human health, should not be exceeded. After that, a secondary standard regarding exposures that threaten vegetation and property, and therefore human welfare, is developed. Finally, the states are required to create plans to bring environmental levels down to the standard, for example by requiring coal-burning power plants to install stack scrubbers that recapture the sulfur dioxide they produce. After all this, the agencies are supposed to police the environment in order to find and prosecute violators—for example, a power plant that emits more sulfur dioxide from its smokestack than the state's plan permits.

This is just the bare bones of what must be done to establish and enforce air pollution standards for *each* pollutant. EPA also publishes "rules"—detailed technical specifications on how to reduce air pollutants to the specified levels. The rules on national air pollution standards are contained in a forty-one-page document that dictates the design and operation of the detection equipment and exactly how local pollutant levels are to be measured. The rules also specify a long list of data that must be acquired from each potential source, for example calling for some eighty separate pieces of information from every power plant in the country. Such rules, analytical procedures, and computations apply to each of the thousands of pollutants that have "an adverse effect on human health or welfare."

All this accomplished (a stage thus far reached for only a fraction of the known pollutants), EPA and the state agencies must then respond to the parties affected by their actions: the companies that object to the cost of the controls they would need to install; the people who may be injured by exposure; industrial, labor, and environmental organizations; the fifty states; other federal agencies. Many of the regulatory actions are challenged in formal hearings and eventually in the courts, often leading to another round of computations, rule making, and standard setting.

In the nearly twenty years of its life EPA has, at great cost, created a monumental technical and administrative apparatus to establish allowable standards, to define the control procedures that are expected to achieve them, and to enforce the resultant regulations. When pollution is prevented, none of this is necessary. And as we have seen, the creation of this elaborate and costly machinery has been in vain. It has produced very little concrete improvement in environmental quality, and step by embarrassing step, enforcement of the laboriously constructed standards has evolved into a distant hope. It hardly requires a sociological survey to determine the response to this failure. It justifies the polluters' inaction and intensifies the public's frustration. It erodes the integrity of regulation and diminishes the public faith in the meaning of environmental legislation. This is a price we pay for failing to attack environmental pollution at its origin.

There is a basic flaw embedded in the U.S. environmental laws: they activate the regulatory system only after a pollutant has contaminated the environment—when it is too late. This fault has spread through the system of environmental regulation, creating practices that seem reasonable and safe from criticism, but which on analysis can often be seen to violate, at their best, common sense and, at their worst, social mores and the public interest.

Consider, for example, the actual effect of the widely approved aim of the national environmental effort: to establish standards of exposure that are supposed to protect human health and the ecosystem. For this purpose, EPA determines some "acceptable" level of environmental harm (for example, a lifetime cancer risk of one in a million) that can presumably be achieved by adhering to the appropriate standards of exposure. Polluters are then expected to respond by introducing control measures that will bring emissions or ambient concentrations to the required levels. If the regulation survives the inevitable challenges from industry (and in recent years from the administration itself), the polluters will invest in the appropriate control systems. Catalytic converters are appended to the cars, and scrubbers to the power plants and trash-burning incinerators. If all goes well—and it frequently does not—at least some areas of the country and some production facilities are then in compliance with the regulation.

The net result is that the "acceptable" pollution level is frozen in place. The industries, having heavily invested in equipment designed to just reach the required level, are unlikely to spend more money on further improvements. And the public is told that the accompanying hazard to health is "acceptable," in the hope that they will be satisfied and stop complaining. Thus, the environmental effort has been based on the social acceptance of some hopefully low risk to health. But this conflicts directly with national policy, as established by the National Environmental Policy Act of 1969 (NEPA), which states as its purpose:

> To declare a national policy which will encourage productive and enjoyable harmony between man and his environment; to promote efforts which *will prevent or eliminate damage to the environment and biosphere* and stimulate the health and welfare of man. [emphasis added]

The words "prevent or eliminate damage to the environment" are unequivocal; they call for eradicating pollution, not reducing it to some "acceptable" level. But NEPA's legislative purpose has been subverted by the very institution that it has empowered. Yet public opinion continues to support the original intent. Thus, in 1986, when a CBS/*New York Times* poll wanted to test the limits of what appeared to be a growing public demand for environmental progress, it confronted its respondents with a deliberately "extreme" position and asked: "Do you agree or disagree with the following statement: Protecting the environment is so important that requirements and standards cannot be too high, and continuing environmental improvements must be made regardless of cost." Sixty-six percent of the respondents agreed, and 27 percent disagreed. The positive response to this "extreme" statement continues to increase, reaching 80 percent in 1989.

There is a sharp contrast between standard setting and the practice of public health. The medical effort to improve public health does not stop at some socially convenient point, but aims at eradicating disease—literally eliminating the risk to health. Public health doctors did not, after all, decide that the smallpox prevention program could quit when the risk reached one in a million. They did not propose an "acceptable" incidence of smallpox as a standard, erecting an administrative barrier against any further improvement. They kept working until the disease was permanently eradicated.

In a way, current environmental practice is a return to the medieval approach to disease, when illness—and death itself—was regarded as a debit on life that must be incurred in payment for Original Sin. Now this philosophy has been recast into a more modern form: some level of pollution and some risk to health are the unavoidable price that must be paid for the material benefits of modern technology. We have an authority no less than *Time* magazine for this new theology. Here is their response to the appalling deaths in the chemical accident at Bhopal, India:

> The citizens of Bhopal lived near the Union Carbide plant because they sought to live there. The plant provided jobs, the pesticide more food. Bhopal was a modern parable of the risks and rewards originally engendered by the Industrial Revolution. . . . There is no avoiding that hazard, and no point in trying; one only trusts that the gods in the machines will give a good deal more than they take away. . . .

Like the medieval priests who accepted the Black Death as the "will of God," *Time* would have us accept the Bhopal catastrophe—where more than two thousand people died and many more were blinded and maimed for life—as the "will of the Machine." In keeping with the times, this new theology is expressed quantitatively, ending in a prayer that economic good will outweigh environmental harm. This is the inevitable result of basing the environmental effort on standards of acceptability. Immediately, the question becomes "Acceptable as compared to what?" and the issue of how to set the standard becomes a battleground for contending economic, political, and moral interests. But these conflicts are elaborately clothed in statistics so that they can masquerade as "science."

Since a standard represents a point on a scale, its practical meaning depends entirely on the nature of the scale. Although the position of the pointer is simply a number and therefore objective, the choice of the scale and therefore the *meaning* of the number is entirely arbitrary. This creates an opportunity to disguise self-interest as science, for the scale is readily manipulated to govern the apparent meaning of the standard. For example, a standard governing environmental exposure to a toxic chemical can readily be placed on a scale of *risk* by comparing the annual probability of dying from that level of exposure with other risks of death. Thus, EPA has estimated the risk of cancer to the general population from exposure to dioxin as 0.000003 percent annually, a purely objective number. But in the hands of the American Industrial Health Council, an industry lobby on cancer policy, the practical significance of such a number is effectively reduced to zero. The council reports that other risks of death are much higher: the risk of a fatality from playing football is 0.004 percent per year; from canoeing, 0.04 percent; from motorcycle racing, 1.8 percent. The council then points out:

> Society has chosen not to prohibit any of these activities, or even activities with much higher risks (e.g., the Indianapolis 500). There are relatively few activities that pose such a high

> risk that society has banned them completely (e.g., going over Niagara Falls in a barrel, or attempting suicide).

This implies that since a risk of 1.8 percent (motorcycle racing) is acceptable, and a nearly 100 percent risk (going over Niagara Falls in a barrel) is not, the proper standard for the risk of cancer from an environmental chemical lies somewhere between. On this scale, the dioxin risk is effectively zero, allowing environmental agencies to invoke the legal canon *de minimis non curat lex*—that the law is not concerned with trifles—to dismiss the dioxin cancer risk as unworthy of any further thought, let alone regulation.

While this approach satisfies industry and industry-oriented regulators—who collaborate in creating and validating the risks—it is much less comforting to the people who bear them. People living downwind from a dioxin-emitting trash incinerator are likely to be concerned about the cancer risk, however small, because unlike the hazards of canoeing or playing football, it is imposed on them involuntarily. They are also likely to ask what benefits accrue to *them* from the incinerator that creates the risk. And this in turn quickly leads to the decisive question: Is there a less risky way of dealing with the trash? (As we shall see in chapter 6, there is.)

In the *Time* prayer, the risk of enduring the chemical catastrophe is balanced by the supposed benefit of producing the chemical. In the recent history of the environmental effort, this "risk-benefit" approach to regulation has achieved the status of a fad, often with peculiar results. A particularly bizarre but eminently logical example was elicited by a decision by the Occupational Safety and Health Administration during the Carter administration to sharply restrict the production and use of a pesticide, DBCP, which had been shown to cause sterility in workers exposed to it. The executive secretary of the National Peach Council, Robert K. Phillips, asserted for the peach growers who wanted to continue using the pesticide, "We do believe in safety in the workplace, but there can be good as well as bad sides to a situation." According to Mr. Phillips, as quoted by the *New York Times:*

> While involuntary sterility caused by a manufactured chemical may be bad, it is not necessarily so. After all, there are many people now paying to have themselves sterilized to assure they will no longer be able to become parents. . . . If possible sterility is the main problem, couldn't workers who were old enough that they no longer wanted to have children accept such positions voluntarily? Or could workers be advised of the situation, and some might volunteer for such work posts as an alternative to planned surgery for a vasectomy or tubal ligation, or as a means of getting around religious bans on birth control . . . ?

The logic is impeccable—and the morality outrageous.

For the most recent twists in the risk/benefit game, we can turn to that eminent source of public morality, the Reagan Office of Management and Budget (OMB). The OMB created a new concept, "risk management," which was adopted as EPA policy in dealing with environmental standards. According to the OMB, "Risk management takes the scientific risk assessment and combines it with other information, such as the cost and feasibility of reducing risks, to determine how to reduce risks most efficiently." For this purpose, the OMB computed the dollar costs of the controls needed to meet a regulatory standard that is supposed to reduce the risk. Then, dividing the annual cost of the regulation by the annual reduction in risk (expressed as the number of lives saved) that it is expected to generate, a "cost per life saved" is computed.

It turns out that certain government regulations save lives very efficiently. For example, regulating unvented space heaters costs $70,000 per life saved, surely a worthwhile investment. But other government regulations have been very wasteful, according to the OMB. Regulating workplace exposure to the carcinogen ethylene oxide costs $60 million per life saved, and the regulation that bans the hormone DES (diethylstilbestrol) in cattle feed (DES has caused mothers exposed to it to give birth to daughters later afflicted with a high incidence of vaginal cancer) costs $132 million per life saved.

Mindful of the taxpayers' dollar, the OMB concludes that

> some activities probably should be regulated more stringently than they are now, and others less. For example, imagine that there are two types of activities that can be regulated—activity A that costs $25 million per statistical life saved and activity B that costs $250,000 per statistical life saved. Spending $100 million less on activity A would increase predicted risk by four deaths. Applying that $100 million to activity B would reduce risk by 400 deaths. Consequently, with no change in total outlays for risk reduction, 396 more lives could be saved.

Of course, if the OMB really wished to put the government in the business of cost-effective lifesaving, there are other, far more reasonable opportunities, especially in the area of cancer. Suppose, for example, that the government wanted to save the 100,000 lives lost each year to cancer from cigarette smoking. And suppose that the government elected to do this by the simple expedient of eliminating the U.S. cigarette industry. In order to avoid hardship, the government would pay the industry $17.3 billion annually to stop producing cigarettes—the amount of its tobacco purchases, wages and other costs, and profits in 1987. It would then cost about $17.3 billion to save 100,000 lives annually, or $173,000 per life—a bargain comparable to the OMB's model of efficiency, regulating unvented space heaters. Obviously, the Reagan OMB was after some other goal.

All such calculations depend, of course, on the dollar value placed on a human life. Economists have developed several ingenious ways of dealing with this intimidating task. The method the OMB prefers is based on "the wage premiums workers require for accepting jobs with increased risk." From such data, various academic studies have estimated that a human life is worth between $500,000 and $2 million. This suggests to the OMB that, while regulating unvented space heaters is cost-effective, regulating DES in cattle feed is not. This judgment carries out the mandate of Executive Order No. 12291, in which President Reagan required that "regulatory objectives shall be chosen to maximize the net benefit to society."

Some economists have proposed that the value of a human life should be based on a person's earning power. It then turns out that a woman's life is worth much less than a man's, and that a black's life is worth much less than a white's. Translated into environmental terms, harm is regarded as small if the people at hazard are poor—an approach that could be used to justify locating heavily polluting operations in poor neighborhoods. This is, in fact, only too common a practice. A recent study shows, for example, that most toxic dumps are located near poor black and Hispanic communities.

Thus, thinly veiled by a facade of seemingly straightforward numerical computation, there is a profound moral or political judgment: that poor people who lack the resources to evade it should be subjected to a more severe environmental burden than rich people. Obviously, this is an issue that ought to be open to public judgment, rather than being stifled in bureaucratic arithmetic. These examples warn us that environmental standards and their manipulation in the name of "risk management" are likely to raise deep-seated, troublesome social issues—which in recent practice have been swiftly reduced to "solutions" by forcing them into the Procrustean bed of the free-market economy.

Another illuminating example of the moral boobytraps built into the cost-benefit philosophy is provided by the EPA's justification for further reducing the lead allowed in gasoline. The EPA carried out a cost-benefit analysis by toting up the cost of more expensive antiknock agents and refinery changes (about $600 million in 1986) and comparing them with the monetary value of the benefits: less engine damage from the lead additives, better fuel economy, and the savings in the cost of diagnosing and treating children with excessive lead in their bodies (a total of about $7 billion in 1986). The EPA was pleased to note that the dollar value of the benefits is more than ten times the cost—clearly a bargain that ought to be snapped up. In these terms, the decision seems to have no more moral content than choosing one brand of breakfast cereal over another. Suppose, however, that a new medical advance were to sharply reduce the cost of treating lead poisoning, bringing the dollar value of the benefits below the cost of controlling lead. What then? Would this new cost-benefit balance justify continued lead pollution?

Thus, the apparently "objective" cost-benefit approach is quickly engulfed in deep moral and political issues: Should society try to mitigate *all* human suffering regardless of cost? On the other hand, since social resources are limited, is it not reasonable to alleviate suffering in keeping with some measure of effectiveness? If a monetary measure is objectionable, what better measure can be used? It is, of course, unreasonable—and probably unwise—to ask the EPA or the OMB to answer these questions, which have long survived much more thoughtful efforts to resolve them. But it is not unreasonable to insist that the issues should be acknowledged rather than reduced to a trivial monetary "solution." By masquerading as objective science, the risk-benefit equation deprives society of the opportunity to confront such fundamental questions. Encumbered by a fundamentally flawed approach to its task, the EPA—goaded by the

OMB—has been driven into positions that seriously diminish the force of social morality. This, too, is a cost of the environmental failure.

Environmental agencies have become remarkably creative about finding new ways to retreat in the face of failure. The latest tactic is the Humpty Dumpty approach. In *Through the Looking Glass,* Alice gets into an argument with Humpty Dumpty, who claims that the word "glory" means "a nice knock-down argument." When Alice objects to this arbitrary redefinition, Humpty Dumpty says:

"When *I* use a word it means just what I choose it to mean."

Alice replies: "The question is whether you *can* make words mean so many different things."

Humpty Dumpty's response is unanswerable: "The question is, which is to be master."

Humpty Dumpty's freewheeling philosophy has begun to take hold in regulatory circles. For example, when tests of fly ash from trash-burning incinerators showed that it was sufficiently contaminated with toxic metals to qualify as a "hazardous substance," the New York State Department of Environmental Conservation issued a remarkable pronouncement: metal-contaminated fly ash is not a hazardous waste, it declared, but a "special waste." Of course, this was not just a silly linguistic exercise. It meant that, unlike an ordinary hazardous substance, fly ash need not be consigned to a heavily protected, expensive landfill—an additional cost that might cripple the incinerator industry.

The EPA and other regulatory agencies have put a great deal of effort into defining a "hazardous waste." Clearly, the public must rely on the integrity of this definition in dealing with Superfund sites (old toxic dumps scheduled for federally funded cleanup) and a whole range of contaminated materials. The linguistic detoxification of fly ash may be a handy expedient for New York State, which soon may be emulated by the EPA as well. (A bill to legitimize the classification of hazardous incinerator ash as "special waste" was introduced in Congress in 1989.) But with Humpty Dumpty in charge, the public has good reason to question who is really the "master" that decides what environmental regulations mean. The loss of public confidence is another price that we pay for the environmental failure.

The OMB Regulatory Program has had a good deal to say not only about the proper balance between environmental risks and costs, but also about the scientific procedures used to estimate the risks, especially of chemically induced cancer. The OMB has been concerned that "risk assessments of health hazards often inform the regulatory officials and the public of only the high end of the range of uncertainty of the risk, i.e., only what the most cautious estimates are."

This reflects the practice, in scientific considerations of health effects such as the risk of cancer from exposure to a certain amount of a carcinogenic chemical, of estimating the most probable "upper bound" of the risk—that is, the greatest risk that is likely to be incurred by exposure. An example of this approach, which the OMB called "overly conservative," is the practice of basing cancer risk estimates on the results of tests with particularly sensitive animals. According to the OMB, this tends to exaggerate the risk to humans, because it assumes that humans are as sensitive as the most sensitive animal tested. The OMB claims that "a more accurate estimate could be derived from a weighted average of all the scientifically valid, available information"—that is, by basing the predicted effect in humans on the *average* sensitivity of test animals.

It is useful to reduce these generalities to practical terms. A good example is the substance AAF (aminoacetyl fluorene), a powerful carcinogen. Fed to rats, AAF invariably induces a high incidence of liver cancer; when fed to guinea pigs, however, the cancer incidence is zero. Using the OMB's "weight-of-the-evidence" approach, this information would lead to the presumed "best estimate" that the sensitivity of people to the carcinogen will lie halfway between zero and the high sensitivity of the rat.

But we actually know a good deal more than that. It turns out that AAF must be acted on by a liver enzyme before it becomes carcinogenic and that rats possess the enzyme, while guinea pigs do not. In fact, this discovery has led to a long series of studies which show that most carcinogens must be chemically transformed in this way before they induce a cancer, and that the activity of the enzyme in people is genetically determined, varying a great deal from one person to the next. This is a typical situation in the human population: a very wide range in the expression of any single genetically controlled characteristic—for example, hair color—resulting from the enormous genetic heterogeneity of the population. In contrast, cancer tests are done on carefully inbred genetic strains of animals of the same age, physiological condition, and usually sex, that vary much less than people. Thus, while all the rats in a cancer test will be albinos with white fur, in people, apart from very few albinos, hair color ranges from platinum blond to black, with many shades in between.

When a factor such as sensitivity to a chemical carcinogen varies from one species or strain of animals to the next, the human population can be expected to include individuals with sensitivities that differ over the whole range of variation—that is, in the case of AAF, the range from from guinea pigs to rats. This is due not only to the genetic heterogeneity of the human population, but also to the fact that—unlike the test animals—the population includes people of different ages and physiological conditions, factors that frequently influence vulnerability to disease.

As a result, when the statistical basis of an exposure standard is chosen, it really represents a decision about what part of the human population is to be protected. If the standard is based on the most carcinogen-sensitive test animal, it will tend to protect the most vulnerable people: infants, the elderly, and people too poor to be healthy, as well as everyone else. On the other hand, if as the OMB prefers, the standard is based on some statistical average, it will protect people of average sensitivity or less—half the population—leaving the other half outside the protection that the standard provides. A standard based on the upper bound risk protects everyone; the OMB's standard cuts the protected population in half.

This is, of course, not a scientific decision, but a moral judgment. The OMB, however, has disguised it in the statistical language of science, with the lofty declaration that it is "a more accurate estimate . . . derived from a weighted average of all the scientifically valid, available information." In contrast, the moral imperative that governs the preventive, public health approach is that the health of all the people—not half of them—should be protected, and that extraordinary measures should be taken to protect those least able to resist disease. That is why vaccination programs are usually targeted on the most vulnerable groups: the very young, the old, and the weak. It explains the worldwide agreement among radiation specialists that radiation exposure limits for the general population should be thirty to a hundred times more stringent than the limits allowed for workers. Thus, the OMB has constructed a pseudoscientific facade that hides a far-reaching moral decision, protecting it from the open public discussion that would surely countermand it. This is also a social cost of the misguided program of environmental regulation.

The integrity of science is the next casualty. It is hardly news that scientific decisions about environmental hazards have consequences that extend far beyond the objective domain of the weekly department seminar. Although the scientific participants may be convinced that their decisions are evenhanded and objective, the consequences are not. Each such decision means that some people will save a good deal of money and others will spend more; that some people will be more concerned about their children's health, and others, less; that the decision will create a political problem for a few people and a welcome political opportunity for a few others. These are simply the facts of regulatory life.

It is perhaps not surprising that the difficulties generated by administering a fundamentally misdirected environmental program should lead bureaucrats into shifty logic and questionable morality. But science is supposed to be at least to some degree immunized against these human frailties. Unfortunately, in the area of environmental science, the immune system has been weakened under outside pressure—for example, from the OMB.

In the last few years, EPA scientific committees, echoing the OMB's call for "less conservative" carcinogen standards, have reviewed earlier standards and have recommended that they should be relaxed. An instructive example is provided by one of the most difficult—and controversial—problems: the cancer risk from exposure to polychlorinated dibenzodioxins and dibenzofurans, which for the sake of simplicity are usually lumped together under the common term "dioxin."

Based on a 1985 assessment of the cancer risk of dioxin to people, the EPA has established a criterion for cleanup of dioxin-contaminated soil that allows no more than 1 part per billion. For Syntex Agribusiness, Inc.—the company liable for the dioxin cleanup costs in Missouri, where an entire dioxin-contaminated town, Times Beach, had to be evacuated—this is a matter of money. This is made explicit in a 1986 technical article by members of the Syntex staff. The article shows that if the standard were relaxed from 1 to 10 parts per billion, cleanup costs would fall by two-thirds. The Syntex paper calls on regulatory agencies such as the EPA to give careful consideration to the "methods and assumptions used in any risk assessment of TCDD-contaminated soil (TCDD is one class of dioxins). The primary reason is

[that] the cost of remediation varies dramatically with the degree of clean-up." The Syntex scientists concluded that the EPA 1985 risk assessment should be sharply reduced. Such a reduction in the cancer risk would have powerful consequences, not only reducing the cost of the cleanup in Missouri and many of the Superfund sites, but also enhancing the environmental acceptability of trash-burning incinerators, weakening the claims of the Vietnam veterans who were exposed to Agent Orange, and affecting the outcome of numerous court cases.

The scientific issues relate to the role that dioxin plays in the biological process that leads from a chemical exposure to the appearance of a tumor. According to a well-known theory, this process is characterized by two sequential steps. First, a substance (an "initiator") induces genetic change in the exposed cells. Then, another substance (a "promoter") causes these now predisposed cells to proliferate and produce a tumor. Mathematical models based on animal experiments have been developed to estimate the risk of cancer in people exposed to given amounts of dioxin. Some models assume that dioxin is an initiator (or a complete carcinogen, capable of both effects), and others assume that it is a promoter. The risks computed from initiator-based models are generally considerably greater than the risks computed from promoter-based models. The 1985 EPA risk assessment assumed that dioxin is not a promoter but a complete carcinogen. Not surprisingly, the Syntex scientists believe that dioxin is a promoter.

At this point, a quick review of the experimental facts is useful. Rats and mice exposed *only* to dioxin exhibit a significant incidence of cancer; this implies that dioxin is a complete carcinogen capable of both initiation and promotion. However, dioxin lacks a diagnostic property of an initiator: it does not cause genetic mutations. Dioxin also sharply increases the carcinogenic effect of a previously applied initiator, thereby seeming to act as a promoter. Yet dioxin also lacks a diagnostic property of a promoter: there is no clear evidence that it causes cell proliferation. According to the actual evidence, then, dioxin is *neither* an initiator (or complete carcinogen) *nor* a promoter. This means that the initiator-promoter scheme is not a sensible way to account for the effect of dioxin on cancer incidence.

Still, there remains the paradoxical fact that rats and mice that are exposed *only* to dioxin exhibit a significant incidence of tumors, in proportion to the dose. Apparently, dioxin acts like a complete carcinogen, but does so in ways not encompassed by the conventional theory. This apparent paradox is explained by the fact that dioxin greatly enhances the activity of the enzyme system that converts most environmental carcinogens into active agents. Apparently, dioxin can so powerfully stimulate the enzyme as to sharply increase the activity of the small amounts of carcinogens present in the laboratory food, water, and air and thereby intensify their effect on tumor incidence. In effect, dioxin influences tumor production by enhancing the activity of carcinogens in the animal (PCB and probably DDT act in the same way).

The group of experts that prepared the 1985 dioxin cancer risk assessment, computing from the effect of measured doses on rats and mice, concluded that dioxin is the most potent cancer-inducing synthetic chemical. Since 1985 this assessment has been used to evaluate the risk of dioxin-induced cancer from the emissions of trash-burning incinerators, from the exposure of Vietnam veterans to the military defoliant Agent Orange, and from the effect of dioxin-contaminated soil in Missouri. Based on the exposure standard derived from the 1985 EPA assessment, it has been found that a number of incinerators generate lifetime cancer risks ten to twenty times greater than the one-in-a-million guideline and that it was necessary to abandon the town of Times Beach, Missouri. Dioxin rapidly became the most highly visible example of the menace of environmental cancer—and a major target of industry publicists anxious to quell what they regard as a dangerous outbreak of "chemophobia."

All this was fair game in the OMB campaign to minimize the significance of environmental carcinogens. Obediently, in 1987 EPA appointed a task force of staff scientists to review the 1985 dioxin assessments "in the light of new data or alternative interpretations." The task force produced a draft report that so slavishly follows the OMB's "scientific" precepts of risk management as to create an illuminating caricature. The report notes that in addition to the 1985 risk assessment, which assumed that dioxin is a complete carcinogen, other agencies have developed assessments nearly 100,000 times lower, based on the assumption that dioxin is a promoter. Confronted by this disparity, most scientists would then examine the relative validity of the two conflicting theories and base their own risk assessment on the one that seemed most plausible—or reject both of them. But the task force was marching to a different drummer—the OMB—and proceeded, in keeping with OMB principles, to "make use of all the data" by averaging the five disparate values. This gambit reduced the 1985 estimate of cancer risk sixteenfold, more than enough to decrease the Syntex Company's financial responsibility for dioxin cleanup by two-thirds.

Perhaps I can best explain this OMB-induced parody of science with a parable. Let us suppose that there is an animal in a closed room, which from previous experience is known to bite, but with an undetermined severity. A scientific task force is called in to evaluate the bite risk. Opinions differ. One group says, "If we assume that the animal is a lion, and apply an appropriate mathematical model, we can compute that the risk from its bite is ten in a million." The other group prefers to assume that the animal is a dog, and its model yields a risk of 0.001 in a million. The scientists engage in a bitter debate, and in an effort to resolve it, they manage to provoke some sound out of the unseen animal. It neither barks like a dog nor roars like a lion. Finally—since the report is due—they reach a consensus: the risk is midway, logarithmically, between 0.001 and 10, or 0.1.

Now, what's wrong with this picture? First, consider the decision to take the midpoint of the two types of risk estimates. Clearly, if the animal is a dog, it should be obvious that it is not a lion—and vice versa. This leads to the conclusion that at least one of the two risk estimates is wrong. But since the animal neither barks nor roars, *both* sides are wrong, which destroys the logical base of both risk estimates—a defect that surely cannot be cured by averaging them.

Eventually, some particularly hardy soul decides to open the door and actually look at the animal. It turns out to be a monkey, which then playfully slides open a door at the back of the room, turning loose a loudly roaring lion. Let me suggest that the monkey—which, unlike the two assumed animals, is *really* in the room—represents the actual observations about the enzyme-enhancing role of dioxin in carcinogenesis (as contrasted with the initiator-promoter assumptions). What dioxin (the monkey) does is to release the serious effects of the complete carcinogen (the lion).

Obviously, the task force report on the dioxin cancer risk fails to meet certain rudimentary requirements of scientific discourse. However, this may not be the proper standard for judging the report. It is not entirely clear that the report is to be regarded as a purely scientific document. The report itself yields a mixed reading on this issue, for it speaks of the conclusion as representing a rational *science policy* position for EPA.

I have emphasized this last phrase because there is an inherent contradiction between "science" and "policy." In the present context, the relevant attributes of science are, first, its demand for rigorous, validated methods that are independent of the results, and second, its objectivity, or the independence of the data and analysis used to reach the results from the interests of those affected by them. In contrast, Webster defines "policy" as "prudence or wisdom in the management of affairs" and "management or procedure based primarily on material interest."

If the new dioxin risk assessment is a purely scientific exercise, then, for the reasons already discussed, its conclusion that the EPA 1985 assessment should be greatly reduced falls under the weight of its seriously inadequate methodology. If it is, instead, a policy document, or some undefined hybrid, then the document fails to meet a different obligation. It fails to specify what "material interest" was intended to govern the outcome of the exercise: the explicit interest of Syntex Agribusiness and other companies in reducing their dioxin cleanup costs; the Office of Management and Budget's often expressed interest in balancing health risks against the cost of achieving them; or the interest of the American people in preventing the growing assault on their environment and their health.

At this writing, the work of the dioxin task force is still under review by the EPA's Science Advisory Board. Such a review is essential, in view of the serious scientific errors that the task force has managed to make in its effort to arrive at what appears to be a preordained conclusion. But the claim that the task force's position represents a rational *policy* position is subject to another kind of review—by the public. Only the public, hopefully with the aid of their elected representatives, ought to determine whether public exposure to dioxin should be governed by the private interest of Syntex, by the political interest of the OMB, or by the public interest in health. Clearly, the task force's procedures compromise the integrity of science. Yet the most serious failing of the task force report is not its specific scientific errors, for they are so blatant that they can be readily dismissed with a light heart. The far more grievous fault is that this flimsy scientific structure has been erected to protect public policy from public scrutiny.

In sum, the original fault embedded in the U.S. system of environmental regulation—that it attempts the inherently futile task of controlling rather than preventing pollution—has itself spawned a series of new faults that erode the integrity of science, the regulatory process, and public policy. The environmental failure has been very costly, not only in money but in burdening the still unsolved environmental crisis with a heavy heritage of poor public policy disguised as science and poor science disguised as policy.

5

REDESIGNING THE TECHNOSPHERE

LIKE ANY EMBATTLED MINORITY, ardent environmentalists are prone to emphasize the singular importance of their special interest in the properties of the natural environment and in the necessity of protecting it. To some of them, thinking about the fact that the environmental crisis results from decisions that, largely after World War II, introduced ecologically unsound technologies of production may seem like an irrelevant digression. The disparate behavior of the ecosphere and the technosphere also creates conceptual problems. Ecosystems are governed by stubborn, largely unalterable natural forces, while systems of production are subject to the much more chaotic processes that govern human choice. Nevertheless, if we wish to pursue the cause of the environmental crisis to its origin, we need to turn from the fairly rigid but harmonious pattern of nature toward the more flexible but uncertain realm of human decisions.

There is a popular myth that in America "the consumer is king," issuing commands for new products that are obediently met by producers. Clearly, that does not explain the universal appearance, about ten years ago, of unsized socks; surely this was not a response to customers whose feet grew from size 10 to size 13 during the sock's lifetime. Rather, it is the *producer's* interests that instigate such changes, based on the expectation that users will at least tolerate them (as in the case of unsized socks, or the plastic egg cartons that have replaced paper ones); or approve of them (as in the case of the larger postwar cars); or have no choice in the matter (as in the case of trucks replacing abandoned rail lines). It is the producer who risks investing in the new product in the expectation that it will succeed in the market. Moreover, some changes in production technology—for example, building a nuclear power plant instead of a conventional one—do not alter the product (electricity) that the consumer buys, but only its price; these changes, too, are initiated by the producer.

In every case, then, the change in production technology, although it interacts with the market and other relevant social factors, is initiated by the producer and is governed by the producer's interests. In the U.S. economy, the motivation that exclusively governs such investment decisions is increased profit and market share. For example, the economic motivation that led the auto industry to decide to manufacture large, powerful cars rather than small ones has been succinctly expressed by Henry Ford II: "Minicars make miniprofits." John Z. DeLorean, a former General Motors executive, has been more explicit: "When we should have been planning switches to smaller, more fuel-efficient, lighter cars in the late 1960s in response to a growing demand in the marketplace, GM management refused because 'we make more money on big cars.' "

Indeed, it cost General Motors only a few hundred dollars more to produce a Cadillac than a Chevrolet; but the large car could be sold for thousands of dollars more than the smaller one. As we have seen, the profit-motivated decision to manufacture larger, heavier cars led to the choice of high-compression engines to power them; then the laws of physics and chemistry took over, leading to the emission of nitrogen oxides, and smog. Smog is a consequence of the automobile industry's devotion to the "bottom line."

Recent changes in the technology of electric power production that have seriously affected environmental quality have also been economically motivated. When they were persuaded—in part by the government and in part by their own propaganda—that nuclear power would be "so cheap that you wouldn't need to meter it," the utilities hastened to build nuclear power plants, creating the monumental hazards of nuclear accidents and the still unresolved problem of radioactive waste. When this hope turned into an illusion, there was a rapid switch to coal-fired plants because they produced cheaper electricity than oil- or nuclear-powered plants—but much more sulfur dioxide. In the same way, the steel industry's response to the demand for less polluting plants—closing rather than improving them—was motivated by the higher profits obtainable from alternative investments in oil and chemicals. Similarly, petrochemicals are typically more profitable than the products that they replace; for example, detergents yield significantly higher profits than soap, which they have largely replaced. It is fair to say that economic motivation—in particular the expectation of greater profits—has impelled the sweeping, ecologically unsound changes in the technology of production that have occurred since World War II.

This conclusion appears to support the view that there is

an inescapable conflict between two socially desirable goals: economic productivity and environmental quality. This apparent conflict has generated a variety of responses. Businessmen warn that if we insist on rigorous pollution limits, plants will be closed and jobs lost—a threat to workers. Conventional economists suggest that the cost of improving environmental quality should be "internalized" and thus become an added cost of production, to be met by higher prices—a burden on consumers. Ecologically minded economists insist, out of respect for the inherent limits of the ecosystem, that we must give up economic growth—a conclusion with terrible consequences for developing countries.

However, the original premise is less clear than these conclusions would suggest. The decisions that governed the introduction of new production technologies after World War II were certainly guided by market forces. In the best of circumstances, these operate to seek out production decisions that maximize profits because they make the most efficient use of the available resources. Yet a major reason for the environmental faults of the postwar technologies of production is that they are wasteful of resources, especially energy. If this is so, alternative systems of production, more closely guided by ecological principles, ought to be more efficient than those devised by the market-guided system. Yet such ecologically sound alternatives to the postwar production technologies have not attracted investment and entered into successful competition with them.

Of course, the overall economic efficiency of the new technologies that have transformed production since World War II—and that have polluted the country—is unmistakable, for the technologies have generated large increases in productivity in every industrialized country. However, the improvement has been unbalanced. While labor productivity—that is, the value of the product yielded per hour of a person's labor—has increased a great deal, the corresponding productivity of capital and natural resources such as energy has declined. For example, the postwar substitution of plastics for leather has displaced an industry characterized by low labor productivity and high capital productivity with an industry remarkable for its high labor productivity and low capital productivity. To generate the same value of product, the plastics industry uses about one-fourth as much labor as the leather industry, but ten times more capital. The disparity between the two industries' energy productivity—that is, the value of product produced by a unit amount of energy—is even greater: the leather industry requires thirty times less energy. These relationships are typical of the postwar production technologies. They have expanded the economy, but at the expense of a trinity of problems: less demand for labor (a factor in unemployment), a greater demand for capital (a factor in inflation), and much more environmental pollution.

The apparent conflict between economic goals and environmental quality was dramatized a few years ago when the administrator of the EPA—William D. Ruckelshaus at the time—offered the people of Tacoma, Washington, where a local copper smelter was emitting illegal amounts of arsenic, a difficult choice. The EPA proposed a regulatory standard that would, they believed, hold the plant's health risk to about one additional case of cancer per year. (As a local commentator put it, "The smelter god could be satisfied with a single human life.") But adhering to that standard would render the plant unprofitable; the required emission controls would be so costly as to force the smelter to close. Mr. Ruckelshaus asked the local residents to choose between increased unemployment and environmental quality. They opted for jobs over environmental quality—a choice that was nullified when Asarco Inc. in 1985, faced with falling copper prices, decided on its own to close the plant after all.

Mr. Ruckelhaus's proposition assumes, of course, that workers are concerned only with their jobs and to that end are prepared to ignore environmental hazards. For a long time, this assumption governed workers' attitudes, reflecting the general view that occupational and environmental hazards are unavoidable risks that must be endured for the sake of economic prosperity. As a result, industrial workers have often served as involuntary environmental guinea pigs, experiencing the earliest and most intense effects of toxic materials that were later detected in the environment. For example, as early as 1933, three years after the first production of PCB in the United States, it was known that workers exposed to it experienced serious medical disorders. Yet it was not until some thirty years later, after billions of pounds of PCB had been produced—and much of it disseminated into the environment—that the chemical's disastrous environmental impact was appreciated and the substance was banned. Now, led by unions such as the Oil, Chemical, and Atomic Workers, who recognize the link between chemical hazards in the workplace and the environment, workers have begun to incorporate environmental demands into their contract negotiations. And, for their part, grass-roots environmental groups have related their demands to the workers', calling for the elimination of toxic materials from production. Today, Mr. Ruckelhaus's proposition is unlikely to be acceptable to either the community or the workers.

As long as the state of the environment is the inadvertent outcome of the play of such short-term economic forces, and depends on controls rather than prevention, it is subject to a kind of "zero-sum game" and can be improved only at some economic cost. If cars and power plants must be equipped with emission controls, their cost inevitably increases the price of travel and electricity; if petrochemical plants are forced to give up their cheap but hazardous toxic dumps, consumers will pay more for detergents, polyester clothing, and the numerous foods now packaged in plastic; if a steel or copper company decides that it is more profitable to close down a plant than to install pollution control equipment, the workers will lose their jobs.

The economic considerations that have governed such decisions are guided by short-term profit maximization. This is not surprising, since it has been widely acknowledged that U.S. corporate decision makers are generally motivated by short-term gains in profit. Corporate managers may feel that they must prove themselves within a few years—setting a very short time span for demonstrating their ability by increasing profits. In a recent *New York Times* article captioned "Something Basic Is Wrong in America," Akio Morita, chairman of the Sony Corporation, criticizes "the American obsession with getting quick results at the expense of long-term considerations." He believes that "the United States government and American companies should shift their time horizons. Investments are needed today that will pay dividends in 10, 20 or 30 years." Mr. Morita's admonition is validated by the vastly superior economic performance of Japan, as compared with the United States, in recent years. It suggests that the economic gains engendered by the post-1950 production technologies have not been enduring ones. The problem is not that corporate managers insist on making a profit—that is, of course, essential if the enterprise is to survive and generate further investments. The difficulty arises from their insistence on governing new investments by the expectation of quick returns that *increase* their annual rate of profit.

Agriculture provides a particularly useful testing ground of the long-term productivity of post–World War II changes in production technology, because there are detailed government statistics about the economic performance of U.S. farms and its relation to the post-1950 changes in production technology. Between 1950 and 1987, total U.S. output of crops and livestock increased by 80 percent. A major influence has been exerted by very specific changes in the technology of production. The use of mechanical equipment remained constant, labor input decreased by 71 percent, the use of seeds and feed increased by 86 percent—and the use of agricultural chemicals (insecticides, herbicides, and fungicides), increased by 484 percent. Thus, the major change in the technology of agricultural production has been the use of less labor and much more agricultural chemicals. The efficiency with which the various inputs generate the farm output is a major determinant of the farm's net returns. This is expressed as productivity—that is, the ratio of output to input. Computed in this way, between 1950 and 1970 the productivity of labor increased by 513 percent, the productivity of machinery increased by 80 percent, the productivity of seeds and feed decreased by 3 percent—and the productivity of agricultural chemicals *decreased by 69 percent.*

The sharp decrease in the productivity of agricultural chemicals is explained by their ecological faults. As increasing amounts of fertilizer are applied to the soil, crop growth reaches a limit and the yield produced per unit of fertilizer falls, reducing economic productivity. At the same time, the excess, unassimilated fertilizer that leaches through the soil pollutes surface and ground water. Continued use of pesticides breeds resistance in the pests, so that progressively more chemical must be used to achieve the same effect, again reducing economic productivity. Thus, both increased stress on the environment and falling economic efficiency are inherent in the present system of chemical agriculture.

In the last few years, U.S. farming has been gripped by a deepening economic crisis, marked by numerous bankruptcies, especially of family farms. Between 1950 and 1987, the net income of the farm sector (corrected for inflation) *decreased* by 32 percent; the farm sector has produced more and earned less. Part of the problem is that the (uninflated) prices paid by farmers for their supplies and services have increased by 228 percent, while the prices they receive for their products have increased by only 146 percent. As agricultural chemicals become progressively more expensive and less efficient in improving crop production, each year farmers must incur a higher debt to buy them, becoming thereby more vulnerable to bankruptcy. The heavy dependence of agricultural technology on chemicals has hurt not only the quality of the environment but the farmers' livelihood as well. If we add to the agricultural balance sheet the as yet unmet costs of the resultant environmental pollution—nitrate-contaminated water supplies, the increased incidence of cancer among farm workers, and the threat to the health of the entire population due to pesticides in foods—even in the not so long term, the postwar "chemical revolution" in agriculture has the appearance of an economic as well as an environmental disaster. Of course, the chemical companies have profited from the farmers' dependence on agricultural chemicals. Acoording to Richard Rhodes, the author of *Farm*, an informative book about the current status of farming, opponents of chemical farming sometimes say that "a farmer is someone who launders money for a chemical company."

Nuclear power provides another example of why there is good reason to question the long-term economic merit of the postwar production technologies. It has been well established that economic development is closely linked to the availability of electric power, for electricity can be applied with nearly perfect efficiency to production processes ranging from tightening a bolt to driving a train. A bountiful supply of electric power is often the engine that drives economic development, so that when nuclear power—the major postwar innovation in generating electricity—was introduced, it was hailed as an economic panacea. But the reality is very different.

The earliest nuclear power plants were built by U.S. utilities because they were judged to be cost-effective in comparison with alternatives such as coal-burning plants. However, this initial economic advantage soon disappeared. Ten years ago, nuclear power was a young, vigorously growing industry. Seventy-two plants were operating, 95 under construction, and 25 on order. Business was booming. Today there are 110 operating plants, but only 12 under construction, most of them not certain to be completed. Over 100 orders have been canceled, and just two are still on the books. There have been no new orders since 1978. Nearly everywhere, the expansion of nuclear power has slowed down or stopped.

Nuclear power has failed as an instrument of economic development, and the reason was clear even before Three Mile Island and Chernobyl. Nuclear power became an economic failure because of the effort to remedy its environmental faults. Costs have increased nearly tenfold, largely because numerous complex, expensive controls and safety devices have been added in order to reduce environmental hazards. Forced by public opposition to respond to their built-in environmental fault—that they create not only electricity but a massive amount of dangerous radioactivity—the industry added costly controls and safety devices until the escalating cost of building nuclear power plants has priced them out of the market.

The ultimate confrontation between the industry and the public occurred over the Shoreham, Long Island, nuclear power plant. This plant, which had increased in cost from an original estimate of $503 million to a final figure of $5.3 billion, was completed to the satisfaction of the federal Nuclear Regulatory Commission and was licensed to operate. In fact, it will never generate electricity. Under growing public opposition, largely based on the gross impracticality of evacuating the affected population in the event of a serious accident, New York State has purchased the plant from its owner, the Long Island Lighting Company (for one dollar, in return for allowing substantial increases in rates), with the intention of dismantling it. The $5.3 billion investment will yield an economic return of zero.

Nuclear power has failed so dramatically because the high costs generated by its environmental hazards have been *internalized* and are therefore directly reflected in its economic productivity. In other industries, the economic impact of such ecological faults, although real, is often less dramatic only because it is not yet reflected in the industry's profit-and-loss columns. The chemical industry is an instructive example. The enormous and increasing environmental hazards generated by this industry are only too well known. The chemical industry is equally famous for its economic success, having grown in the United States, for example, to a nearly $200 billion industry (in assets), especially because of the production of synthetic organic chemicals since World War II. What is less well known is that a significant effort to rectify the industry's environmental defects would seriously damage its economic viability.

In 1986 the annual output of the chemical industry, as represented by the top fifty products, amounted to 539 billion pounds. Based on the EPA's Toxic Release Inventory, the U.S. Congress's Office of Technology Assessment has estimated that about 400 billion pounds of toxic chemicals are discharged into the environment annually. Some of these are chemical industry wastes and others are the industry's actual products, such as dry-cleaning solvents, that are discarded after being used. In either case, they contribute to the enormous toxic burden imposed on the environment by the chemical industry.

The toxic chemicals enter the environment in various ways: emissions into the air; effluents discharged into sewer systems or directly into surface waters; deep-well injections; surface lagoons; or storage dumps. Only about 1 percent of the industry's toxic waste is destroyed, which is the only way to ensure that these substances, many of them highly dangerous and long-lasting, do not threaten living things. If the present (and still environmentally unsatisfactory) method of destruction—incineration—were applied to the toxic chemicals now discharged into the environment, at an average cost of perhaps $100 per ton, the total annual cost would amount to $20 billion. In 1986 the chemical industry's total after-tax profit was $2.6 billion. The arithmetic is deadly: if the chemical industry were required to eliminate toxic discharges into the environment, the cost would render the industry grossly unprofitable. In effect, the chemical industry is profitable only because it has thus far managed to avoid paying its environmental bill. Proper treatment of its waste, which, given rising public concern over this issue, is likely to be forced on the industry in the near future, will mean higher prices and serious competition from the natural products that it has replaced.

The environmental costs of other industries are more difficult to evaluate, but many of them may also be large enough to drastically unbalance the industries' books when the time comes, as it surely will, to pay the bill. How profitable would the power industry be if it were required to eliminate completely its contribution to acid rain—let alone pay for the costs already incurred by this pollutant? What would remain of the auto industry's short-term profits if they were diminished by the need to invest in building a smog-free car?

In general, the short-term profits of the new postwar industries have been high as long as they have not yet felt the impact of their ecological transgressions, which emerge only after an industry has been operating for a time. It is often public concern with the environmental faults that limits the continued economic development of such production technologies. The development of nuclear power in the United States, and in other countries such as Italy and, most recently, the Soviet Union, has been hindered because the public refuses to accept it at any price. Under public pressure, small but similar limitations on the continued economic growth of the petrochemical industry are beginning to develop in the United States: laws to ban the use of disposable plastic food packaging; demands for the banning of particularly hazardous chemicals; local ordinances that favor ecologically benign biological pest controls over chemical ones.

Thus, we have been relying on production technologies that, despite their initial profitability, are limited in their ability to support long-term economic development largely because of their harmful impact on the environment. They represent investments that, guided by the principles of free enterprise, promised to yield the greatest return in the shortest time. Since the social interest in environmental quality or long-term economic efficiency is not taken into account in such decisions, it is not surprising that neither of these socially desirable results have been achieved by these investments.

Another reason for the long-term economic vulnerability of the post–World War II production systems is their heavy reliance on fossil fuels: oil, natural gas, and coal. These fuels are *nonrenewable*—that is, they comprise stores laid down once in the Earth's history, which are therefore irreplaceable. During the energy crisis of 1973 when this fact was popularized, the problem engendered by our reliance on fossil fuels was usually expressed in warnings that "we are running out of oil." However, the real problem with the nonrenewability of oil is not geological but economic.

Oil deposits vary a great deal in their accessibility: some deposits are large and highly productive with little effort, while others are small and yield much less oil for the same effort; some deposits are shallow and less costly to reach, while others are deep or under water and hence more expensive to produce. It is only natural that the most accessible and cheaply produced deposits of oil should be used first. In effect, every barrel of oil taken out of the ground automatically makes the next barrel more expensive.

This theoretical expectation is borne out in practice: the constant-dollar cost of producing oil in the United States, for example, has risen exponentially since the mid-1960s, doubling approximately every 15 years. The problem is not that we are "running out of oil"—that has been true, after all, since the first oil well was drilled in 1859—but that now, after 130 years of oil production, the exponential cost curve is rising rapidly. As a result, as long as we continue to rely on nonrenewable fuels, especially oil and natural gas, a progressively larger fraction of the economic system's output must be invested in producing energy. Yet the only function that energy performs is to provide the work that drives all productive activities—which in turn generate the goods and services that constitute the source of economic wealth.

Thus, because the present energy system is almost entirely based on nonrenewable fossil fuels, it cannibalizes the very economic system that it is supposed to support. The economic difficulties that accompanied the extraordinary rise in the world price of oil in 1973 and 1982 due to OPEC manipulation warned us of the serious impact of this process. Projections based on the exponentially rising cost of producing oil, especially in long-exploited areas such as the United States, indicate that in 2020, without any OPEC interventions, the price of oil—and with it, the price of energy generally—will reach a level, relative to the general economy, comparable to the price in 1973. Thereafter, the exponentially rising cost of producing oil will demand a progressively larger share of economic output, creating a permanent energy crisis that can only become worse with time. In the long term, this process will erode the economic productivity of energy-intensive industries, especially transportation, electric power, and petrochemicals.

Finally, even in the near term the productivity of the country's economic system has been damaged by our neglect of public environmental facilities such as water and sewage treatment plants. In recent years, especially under the Reagan administration, there has been a sharp decline in financial support for these essential, environmentally important services. The deplorable condition of Boston Harbor—heavily polluted because of grossly inadequate sewage treatment—became a prize example of political irony in the 1988 presidential campaign. Mr. Bush, the "environmental president," attacked Mr. Dukakis over this issue, although it was Reagan's fiscal policies that were responsible for it.

The serious *economic* consequences of this environmental neglect are evident in a recent report from the Chicago Federal Reserve Bank entitled *Is Public Expenditure Productive?* The study was designed to explain what the report calls "the productivity slowdown in the last 15 years" (i.e., 1970–85). During that period of time, the rate of increase in total productivity—that is, the efficiency with which the production factors (labor, capital, and resources) contribute to economic output—declined sharply in the United States. The growth rate in productivity, generally regarded as the best measure of an economy's strength, averaged 2.2 percent per year from 1950 to 1970, but fell to 0.6 percent during 1971–85; it was only 0.2 percent in 1980–85. Seeking to explain this very troublesome trend, the report used statistical analysis to show that it can be largely accounted for by a parallel decline in public investment in the national infrastructure, such as sewer and water systems, mass transit, roads, and bridges. The growth rate of investments in such "public capital" fell from an average of 4.1 percent per year in 1950–70 to 1.6 percent in 1971–85. This analysis shows that the strength of the U.S. economy, which has been in a serious decline, depends critically on these essential social investments, which have a major impact on environmental quality.

According to the report, this relationship helps to explain why the United States has the distinction of exhibiting the slowest growth of productivity among the "Group of Seven" countries that comprise the annual economic summits of recent years (United States, Canada, United Kingdom,

France, Italy, West Germany, and Japan, listed in order of increasing annual growth in productivity). Among these nations, there is a statistically significant relationship between their rate of growth in productivity and their relative investment in public facilities, expressed as a percentage of GNP. Thus, during 1973–1985, Japan's rate of improvement in productivity was more than five times greater than the U.S. rate—and the percent of its GNP invested in "public capital" was 17 percent higher than the U.S. figure.

The main conclusion that can be derived from the Chicago Federal Reserve Bank study is fairly simple: environmental pollution is bad for business. A local expression of this precept is the report from New York City that fish consumption declined by about 50 percent because of the debris from inadequate sewage treatment facilities that washed up on their beaches in the summer of 1988.

In sum, the conflict between economic productivity and environmental quality is apparent only if economic gain is defined as short-term profit, and if environmental improvement is restricted to futile but costly efforts to control pollution rather than prevent it. On the other hand, if one takes a more fundamental, preventive approach to the problem of environmental quality by recognizing that it is inherent in the design of production technology, it is possible to find ways of improving *both* the environment and the economy.

The electric power industry provides an informative example. This industry is a notorious source of pollution, responsible for a great deal of dust, sulfur dioxide, and nitrogen oxide, contributing considerably to acid rain and generating the threat of nuclear disasters. The industry is also in a precarious economic condition; the utilities' heavy investment in huge power plants—especially nuclear plants—has driven the cost of producing electricity sharply upward and has threatened some of the companies' economic stability. (A 1984 *Business Week* cover story asks the question "Are Utilities Obsolete?," citing the view of industry experts that "the large central power station is an idea whose time passed a decade ago.") A major reason for these economic difficulties is that the present technology of electric power production is highly centralized. Each new plant is very large because it is designed to provide the needed capacity well into the future. It therefore requires a huge capital investment, so that when a new plant begins to operate, the system inevitably has excess capacity and for a time part of the capital investment yields little or no return. (Some return can be obtained by selling the excess power to another utility, but it is much less than the return from direct sales.) Moreover, the transmission system that distributes electricity from central stations is costly and consumes a significant fraction of the power. Finally, centralized power plants are inherently inefficient because, for inescapable thermodynamic reasons, two-thirds of the energy available from the fuel is discarded into the environment as waste heat. This means that the industry wastes two-thirds of the fuel that it uses and causes three times as much pollution per unit of useful energy produced than it would if the wasted energy could be usefully recaptured.

The heat ordinarily discarded by a central power station can readily be recaptured and used, for example to heat homes; such a plant is known as a cogenerator. But, except in densely populated areas, large centralized plants cannot be used in this way, because heat can be effectively transmitted only over short distances. No one wants to live that close to a power plant, especially if it is nuclear. (Some people are in this unhappy position. Nuclear power plants have been built to provide heat to two Soviet cities, Gorki and Voronezh; the plants must be inside these cities or certainly quite close to them—a situation that must be the cause of a good deal of concern after the Chernobyl disaster.) By redesigning the technology of electric power production in keeping with sensible, ecological principles, both environmental and economic improvement can be achieved. This can be done by decentralizing the production of electricity and installing cogenerator plants just large enough to meet the local demand for both heat and electricity. Such cogenerators can readily be installed in urban multifamily residences; since they are best fueled by natural gas, they create no more environmental impact than present natural gas heating systems, which is relatively low. With the cogenerator operating at a much higher level of both economic and thermodynamic efficiency than a conventional power plant, fuel consumption is reduced, decreasing both environmental impact and the cost of energy—a net gain for both the economy and the environment.

Decentralization of the electric power system is a good preparation for the entry of solar electricity, which is also most cost-effective when produced at the point of use. In contrast with nuclear or fossil-fueled power plants, such a source, for example photovoltaic cells that convert sunlight directly into electricity, has little or no impact on the environment. Moreover, because of the escalating cost of producing nonrenewable energy, sooner or later solar sources become less costly than electricity produced from fossil fuels. Photovoltaic electricity is already cheaper than conventional power sources for remote locations. As the cost of conventional sources rises and that of photovoltaic cells declines, solar electricity will emerge—probably within the next ten years—as a major means of stabilizing the rising cost of electricity.

Similar environmental and economic gains could be made by reorganizing agricultural production along ecologically sound lines. Ecologically, agriculture is a highly effective means of converting solar energy into food and fiber. Given sufficient water, and properly managed, the system can operate provided with nothing more than sunshine. But modern agricultural technology has disrupted this efficient relationship. Now, instead of essential nitrogen being derived from the air by nitrogen-fixing plants, it is supplied in the form of chemical fertilizer, a major uncontrollable source of water pollution. Now, instead of relying on their natural predators to keep insects in check, or on cultivation to control weeds, large amounts of synthetic chemicals—dangerous to the farmer and the environment—are sprayed on the land. Now the fuel needed to drive farm machinery, which could readily be obtained from crops in the form of ethanol (a renewable solar fuel), is supplied by petroleum, a nonrenewable, progressively more costly source of energy. By avoiding these ecologically unsound practices, farmers could actually improve their economic position.

Twenty years ago, this was the view of only a very small minority of seemingly eccentric proponents of "organic farming." Their claims of success were anecdotal and difficult to verify. In 1974, the Center for the Biology of Natural Systems (CBNS), then in St. Louis, began a five-year study to evaluate the performance of conventional and organic farms. Fourteen organic farms in the Midwest were located. These were large farms (400 to 500 acres) that raised the customary crops (chiefly corn and soybeans) using conventional machinery—but they used no chemical fertilizer or pesticides. For the purposes of the study, each farm was matched with a neighboring, nearly identical conventional farm. A five-year comparison of their crop and financial records showed that the organic farms' crop output was 8.5 percent less than the conventional farms'. However, the resultant loss in income was exactly balanced by the savings gained by not buying agricultural chemicals. As a result, averaged over the five-year period, the organic farms produced the same net income per acre as the conventional ones.

Since this initial study, organic farming has spread considerably, although still a small part of total agriculture. In 1989 a study done for the U.S. Department of Agriculture by the

National Research Council of the National Academy of Sciences confirmed the earlier results, concluding from an analysis of fourteen organic farms ranging from Florida to California that they performed economically at least as well as average conventional farms. Organic farms exemplify how the principles of pollution prevention can be applied to agriculture, reducing the environmental emissions due to nitrogen fertilizer and synthetic pesticides to zero, yielding crops free of toxic residues, and succeeding economically as well.

Ecologically sound agricultural practice would also include on-farm production of energy, in the form of ethanol made by fermenting crops and crop residues. A 1982 CBNS study, done for the U.S. Department of Energy, showed, from a computer analysis, that this could be done without reducing food production. The study showed that by substituting sugar beets for soybeans, a typical midwestern farm that raised crops chiefly for animal feed would produce about as much protein as before but nearly twice as much carbohydrate. As a result, the extra carbohydrate could be converted to ethanol, leaving enough carbohydrate and protein to support as much animal production as before. The study showed that after fifteen years, such a crop system would double the farmers' profit, compared to profit from conventional agriculture. Applied to U.S. agriculture as a whole, this new system of agricultural technology could produce enough ethanol to replace about 20 percent of the national demand for gasoline without reducing the overall supply of food or significantly affecting its price.

Here, then, is a way of both improving the environmental impact of agriculture and expanding its economic output. This approach could readily be combined with a sharp reduction in the use of fertilizers and pesticides, for any resultant loss in crop output would be far outweighed by the added production of ethanol. In sum, in agriculture it is possible to achieve economic growth *and* improve the environment—by sensibly applying the principles of ecology.

In the automobile industry, ecologically mandated redesign is also possible. Suppose that, guided by the few environmental successes, we seek to control automotive smog at its origin, the *production* of nitrogen oxides. The goal would be zero production of nitrogen oxides by cars and the complete elimination of this dominant source of smog.

Is this approach really practical? Can smogless engines that do not produce nitrogen oxides be built? They can; indeed, they have been. Every pre–World War II car was driven by such an engine; that is why the country was then free of smog. In fact, nitrogen oxide production can be largely prevented even without giving up the American car's precious overpowered engine (which would still be a good idea). The so-called "stratified charge" engine can do just that. According to a 1974 National Science Foundation study, prototypes were then operating in Detroit, and tests showed that the engine would meet the 90 percent reduction in nitrogen oxide emissions required by the Clean Air Act Amendments. But according to the NSF report, the conventional engine would need to be considerably redesigned, needing a new fuel injector, fuel pump, ignition spark-plug system, cylinder head, piston, intake, and exhaust manifolds. Unlike the addition of a catalytic converter to the exhaust system of the existing engine, this would mean extensive retooling in the manufacturing plants. According to the report, had the auto industry decided in 1975 to take this course, the stratified charge engine could now be driving most U.S. cars—and automotive nitrogen oxide emissions would have been sharply reduced instead of increasing.

But American manufacturers have thus far been reluctant to make the large-scale production changes needed to take advantage of this opportunity. Apart from a limited flirtation with small cars during recent energy shortages, the industry has refused to build cars light enough to be driven by low-compression engines. Nitrogen oxides from automotive transportation could also be eliminated by the substitution of electric motors, which produce no environmental emissions at all, for the present ones. In diesel engines, fuel combustion produces carcinogenic chemicals, and these adhere to the tiny carbon particles emitted by the engine. Here, too, the auto industry has been unwilling to make the considerable changes in engine technology needed to reduce the hazardous emissions.

Production technology encompasses not only the design of a particular facility, such as a car, but the overall system in which it operates, in this case transportation. While remedying the vehicle's inherent faults is useful, much more environmental and economic improvement can be accomplished by addressing the system of transportation as a whole. Compared with railroads, cars are a very inefficient means of long-distance and commuter travel. (They use much more fuel, and therefore produce that much more pollution, per passenger mile traveled.) But they are essential, even with effective mass transit, for some urban and suburban travel. Designed exclusively for this purpose, cars could be much lighter and driven by less powerful low-compression engines or, for that matter, by totally nonpolluting electric motors. Together with the expansion of urban and suburban mass transit systems and the proper design of the electric power plants that serve them, this would virtually eliminate urban ozone and smog. Moreover, lower gasoline consumption would greatly reduce carbon monoxide emissions even without exhaust controls. In the same way, carcinogenic diesel exhaust could be sharply reduced by shifting from truck-borne freight to railroad freight, which is four times more fuel-efficient. While these changes might mean lower profits for the auto industry, by eliminating the mounting costs of pollution they would in the long term represent a net economic gain for the nation as a whole.

In sum, there are ways of attacking pollution at its source—and thereby preventing it—many of which satisfy the social interest in both environmental quality and long-term economic productivity. Had American farmers begun sooner to shift to organic farming and ethanol production, nitrate water pollution would now be falling instead of increasing, the rising level of pesticide pollution would be checked, and farmers would be prospering. If the auto industry had continued, after World War II, to produce small cars; if the railroads and mass transit were expanded; if the electric power system were decentralized and increasingly based on cogenerators and solar sources; if the pitifully small percentage of American homes that have been weatherized was increased—both air pollution and fuel costs could be sharply reduced. If brewers were forbidden to put plastic nooses on six-packs of beer; if supermarkets were not allowed to wrap polyvinyl chloride film around everything in sight and then stuff it into a plastic carrying bag; if McDonald's could rediscover the paper plate; if the use of plastics was limited to those products for which they are really needed, say, artificial hearts or videotape—then we could push back the petrochemical industry's toxic invasion of the biosphere and reduce the escalating cost of waste disposal.

These changes would reverse the present illogical relations among the environment, the system of production, and the economy. At present, economic considerations—in particular, the private desire for maximizing short-term profits—govern the choice of productive technology, which in turn determines its environmental impact, generally for the worse. Logically, however, environmental constraints ought to determine the choice of production technology, and that choice should govern economic investment policy. Since environmental quality has been adopted as public policy, it follows that the hitherto wholly private decisions that determine production technology must become subject to social governance. Logic, applied in this way, would enhance both environmental quality and long-term economic productivity.

6

PREVENTING THE TRASH CRISIS

I HAVE ARGUED, thus far, that the remedy for our spectacular failure to clean up the environment is public participation in the until now private decisions about how goods and services are produced. This is, of course, easier said than done. The view that the nation's welfare depends on "private enterprise" is so deeply embedded in American political life that even to raise the question of possible public intervention is often an open invitation to ridicule. It is nevertheless a fact that recent environmental campaigns have begun to encroach on this taboo, although participants may not have been consciously aware of it.

In one major area of production—nuclear power—public intervention has already had a powerful effect: in the United States it has brought the industry to an ignominious halt.

The nuclear power industry is paralyzed because intense public opposition has made the industry pay its environmental bill, most dramatically by forcing the abandonment of the $5.3 billion plant at Shoreham, Long Island. On the surface, this event dramatizes the power of public protest to stop an environmentally hazardous project, nullifying a huge investment in money and effort. More fundamentally, but unintentionally, it demonstrates how social intervention can govern the choice of production technology—in this case, the means of producing electricity. But this process has been incomplete. Although public opinion countermanded the Long Island Lighting Company's private decision to use nuclear power to produce the electricity that its customers will surely need, this has not resolved the issue of what other source of power will be used to meet that demand. We can turn to another issue—trash disposal—for an example of how social forces, also impelled by environmental concerns, can participate in the actual technological choice. The issue is therefore worth examining in some detail.

Trash is a pollutant that is the inevitable end result of the linear production processes that supply households and commercial establishments with their needed goods. Most of the goods that enter a household sooner or later leave it—as trash. The daily newspaper becomes trash as soon as it is read; the supermarket's shopping bag is converted into trash the moment it is emptied; within a day or two the remnants of the meat and fresh fruit and vegetables it once contained also become trash; and not much later the same is true of the containers that once held milk, pickles, TV dinners, or liquid soap. Less transient household goods—clothes, utensils, appliances, books—may be retained for years before being discarded, but the packaging in which they arrive is quickly consigned to the trash. Households—and commercial establishments as well—are at the receiving end of a process that, along with the desired goods, also delivers materials that in the act of consumption become trash.

The nature of household trash has changed dramatically in the last few decades. When milk was delivered to the doorstep—and the empty bottles were taken away—the only trash generated by milk consumption was the paper bottle cap. Now drinking the same quart of milk leaves behind a container made of layered cardboard and plastic that has no redeemable value and must be burned or buried. Hardware stores once sold screws out of a bin and they were carried home in a small paper bag or one's pocket. Now the same purchase contributes to the household trash bin an unnecessarily large sheet of cardboard and a plastic overlay. When diapers were washed at home or sent to a laundry service, tending the baby generated little or no trash. Now, with disposable diapers in vogue, the same baby produces a huge amount of trash—four pounds or more of soiled diapers per day. It is not surprising, then, that the per capita production of trash has increased by about 40 percent since 1960.

Such statistics encourage the view that the generation of trash is the consumer's fault, reflecting a natural preference for conveniences such as TV dinners and disposable diapers. In fact, most if not all of the changes that have enlarged the trash stream, although often designed to satisfy some real, imagined, or advertising-driven consumer demand, are created, like pollution generally, by producers' decisions.

Beer provides an informative example. Between 1950 and 1970, the number of beer bottles used in the United States increased by more than sixfold, adding considerably to the trash stream. What caused this new trash problem was the breweries' decision to shift from returnable deposit bottles to no-deposit throwaways. In turn, this decision was motivated by the economics of beer production. The beer industry, once a local affair with a dozen or so breweries competing within each city or region, became heavily centralized in the 1950s. Thousands of local breweries went out of business, replaced by a handful of companies operating huge plants, each servicing a large part of the country. Shipping empty bottles back to the brewery over long distances made no economic sense to the large brewers—and the throwaway beer bottle was introduced.

We can safely conclude that this change was not a response to consumers too lazy to carry returnable bottles back to the store. If that were so, the local breweries could have made the change just as well as the new, centralized ones. On the other hand, the disposable diaper has clearly satisfied a strong consumer interest, having captured, since its introduction, most of the market previously held by laundries offering a significantly cheaper service. The plastic shopping bag, a much more difficult trash disposal problem than the paper bag that it has displaced, was motivated by storekeepers' interest rather than their customers': it costs them less.

One reason why, until recently, householders have not reacted to the noticeable increase in trash generation that has accompanied their purchases is that in most places trash disposal is a community responsibility. In cities, for example, each of us is free to set out at the curb as much trash as we have accumulated, confident that the municipal sanitation workers will haul it away. While the rising per capita generation of trash has increased disposal costs—and taxes—the problem that is now recognized as the "trash crisis" arose from a resource infinitely less renewable than taxpayers' cash: land.

For a long time, especially in rural areas, trash was simply piled up at the town dump. Later, in order to cover the malodorous mess, holes were dug and soil was layered over the trash periodically, creating what was euphemistically called a "sanitary landfill." But a hole is, after all, a nonrenewable resource: it fills up. Even disposing of trash in this simple way becomes more difficult as the hole fills and then becomes a mound, until, as it grows, trucks are unable to climb the progressively steeper slopes and the landfill must be abandoned. Then, since the trash keeps coming, a new, more distant, more expensive landfill must be found. In this way, like any other nonrenewable resource, landfills become progressively more costly simply by being used.

Apart from this basic fault, landfills are serious sources of environmental pollution. Often poorly controlled, they are likely to receive wastes—unwanted pesticides and other chemicals, waste motor oil, used cleaning fluids and solvents—that readily leach out of the landfill into underground water supplies and nearby surface waters. Moreover, the landfill's organic waste putrefies and ferments, producing inflammable methane and other gases, some of them quite noxious, that pollute the surrounding air. Thus, there is an inevitable conflict between the landfill's unhappy and inherently limited lifetime and the incessant, increasing flow of trash that it must accommodate. The problem became acute in the 1980s, especially in heavily populated areas such as the East Coast, and the cost of depositing trash in a landfill, the "tipping fee," began to rise rapidly. The national average tipping fee, which had been relatively constant until 1984, more than doubled in the next four years; fees in the Northeast were then nearly twice the national average.

Understandably, municipal officials began to look for a way out of this progressively more difficult situation. It appeared in the form of an old idea: burning the trash. This was a practice frequently employed in a futile effort to reduce the nasty impact of decaying dumps. In the 1950s and 1960s, a number of U.S. cities built trash-burning incinerators, and apartment houses used them to burn their own residents' trash. But the Clean Air Act Amendments of 1970 established new emission standards that the incinerators could not meet; a few added emission control devices, but most were closed down.

Again, the relentless flow of trash continued to burden the landfills' diminishing capacity. Then late in the 1970s, municipal officials were visited by salesmen for a new type of incinerator, now graced with a disingenuous name: "resource recovery plant." (The resource recovered was steam or electricity, produced from the heat generated by burning the trash.) This development was not so much a response to the trash problem as to a problem in another sector of the economy—electric power production. With the collapse of the U.S. nuclear power industry in the late 1970s and with excess capacity in most electric utilities, a number of large corporations were faced with the prospect of losing heavy investments in their power plant manufacturing facilities. To make up for canceled power plant contracts, some of them, including the "Big Four" of nuclear power (Westinghouse, Babcock & Wilcox, Bechtel, and Combustion Engineering) decided to sell trash-burning incinerators instead. The *Wall Street Journal*, in reporting sharp budget cuts at the Westinghouse nuclear power unit, commented that "the company's entry into several new energy-related businesses, notably the burgeoning waste-to-energy field, should soften somewhat the nuclear business' decline." The urgent need to find orders for nearly idle power plant manufacturing facilities generated high-pressure sales efforts. David L. Sokol, president of Ogden-Martin, an incinerator company not associated with the power plant industry, has described the situation this way:

> Late in the Seventies, intrigued by synergistic opportunities, new companies entered the [incinerator] field. Many were equipment vendors trying to increase boiler sales; or engineering/construction services firms, seeking lucrative design

> contracts; or other vendors. An overabundance of electrical capacity, leading to a decline in new utility power plant projects, precipitated their aggressive entry into refuse-to-energy.

The sales campaign was successful. Touting incinerators as "proven technology" and the only alternative to increasingly costly landfills, between 1983 and 1987 power plant manufacturers and several independent companies sold 173 incinerators, costing an average of about $100 million each. A symptom of the cozy relations that developed between city officials and the incinerator industry was the creation, by the U.S. Conference of Mayors, of a subsidiary, the National Resource Recovery Association, which promotes incinerators. Since landfills accounted for 90 percent of trash disposal in the early 1980s, the industry appeared to be on the verge of a huge boom, reminiscent of the early days of nuclear power.

But the curious link between nuclear power plants and trash-burning incinerators persisted. Just as nuclear power failed because it *created* an environmental hazard—radiation—so incinerators turned out to be gravely hampered by the same sort of self-generated environmental hazard, in this case dioxin. In the late 1970s, several technical reports about the stack emissions of European trash-burning incinerators noted the presence of the best-known (and believed to be the most toxic) of the family of 210 compounds commonly called dioxins—2,3,7,8-tetrachlorodibenzo-p-dioxin, usually abbreviated as 2,3,7,8-TCDD. Not much attention was paid to this information in the United States until an incinerator in Hempstead, Long Island, was tested in early 1980 to determine why its stack emissions were so malodorous. There was no reason to blame dioxin for this problem, but, because a chemical related to it had been found in the emissions, samples were sent to a laboratory at Wright State

University in Dayton, Ohio, that specializes in dioxin tests. They contained significant amounts of dioxin. An intense controversy erupted, first among technicians about the validity of the results, and later in the community about their significance. Finally, concerns about the hazard and about technical problems in operating the plant led to its closing. By then a number of cities, among them New York, had decided to build "resource recovery" plants to divert the endless stream of trash from their rapidly filling—or rather, mounting—landfills. The New York City Department of Sanitation (DOS) decided to build the first of eight planned incinerators—a plant designed to burn 3,000 tons of trash per day—at a site in the Brooklyn Navy Yard that the U.S. Navy had ceded to the city.

The DOS plan was opposed by the people who live in the residential community adjacent to the Navy Yard, Williamsburg. At first, they objected to the heavy truck traffic that the incinerator would generate; the DOS responded by agreeing that the trash would be carried to the incinerator by barge. But then word came from Hempstead about dioxin emissions, creating in the minds of the residents a new concern about the health effects of incinerator emissions. They were poorly prepared to deal with the problem. Dioxin had been detected as a highly toxic impurity in chlorinated herbicides such as 2,4,5-T in the 1960s. But it was unknown as an environmental pollutant until 1973, when it was discovered in fish contaminated, during the war in Vietnam, with the defoliant Agent Orange. In 1977, following a chemical plant explosion in Seveso, Italy, that spread only a few pounds of dioxin over a nearby neighborhood but nevertheless required evacuation of the area, the extraordinary toxicity of dioxin was widely appreciated, if still poorly understood.

When Williamsburg residents raised questions about the dioxin hazard, the DOS was ill prepared to answer. The initial response was a blanket assertion that there would be no adverse health effects of any kind from incinerator operations. Then, in August 1983, when the issue of dioxin emissions had become unavoidable, the commissioner of sanitation responded by asserting (in a *New York Times* op-ed article) that "unsorted garbage can be burned without producing dioxins."

Such assertions raised problems for the residents, who had heard, for example, that the Hempstead incinerator emitted dioxin from its stack. Desiring explanations, they sought expert advice. The previously mentioned Center for the Biology of Natural Systems, which I direct, is a research institute that has, as one of its purposes, helping communities solve environmental and energy problems. Accordingly, we responded by accepting the residents' invitations to informational meetings, at which DOS representatives were heard as well. In response to the DOS claim that incinerators do not produce dioxin, we produced contrary evidence: a series of reports from European incinerators showing that each of them produced dioxin in their emissions. Confronted with this evidence, the DOS retreated to a new position: the incinerator furnace would be hot enough to destroy the dioxin. But this contradicted the fact that tests of various incinerators showed that dioxin was emitted even at such elevated furnace temperatures.

The inconsistencies in these claims only reinforced the residents' protests, and whenever the proposed incinerator was considered—hearings were held before the New York City Board of Estimate, several Community Boards, and a Citizens' Advisory Committee set up by DOS—dioxin dominated the discussion. It became clear that the public acceptance of the proposed incinerator would stand or fall on the expected effect of the dioxin emissions on the people exposed to them.

Such a risk assessment calls for a series of technical evaluations, beginning with the assumed rate at which dioxin will be emitted from the incinerator's stack. Then it is necessary to work out how the emitted material will spread downwind from the stack, in particular the amount to which a person will be exposed at the point where the concentration is highest. With this information in hand, the amount of the environmental dioxin that is likely to enter such a person's body is computed. Finally, by comparing that number with the rate of cancer incidence observed in laboratory animals exposed to measured amounts of dioxin, the risk of cancer to a person exposed to the highest expected concentration over a seventy-year lifetime is estimated. This "maximum lifetime cancer risk" is generally used as a measure of the health hazard from the incinerator's dioxin emissions.

These evaluations call upon an extraordinary range of scientific disciplines: the chemistry of combustion; the physics of air movement and dust settling (part of the emitted dioxin is attached to fine dust particles, "fly ash," produced in the incinerator combustion chamber); the movement of dioxin through the food chain; the physiological mechanisms that can carry airborne dioxin into the body (inhalation, ingestion, and absorption through the skin); the biochemistry of the cancer process (at best a poorly understood area of science); the influence of each of the 210 related substances commonly called "dioxin" on the cancer process, as determined from animal tests; determination, from the measured effects on cancer incidence in laboratory animals, of the expected occurrence of cancer in the exposed population.

The DOS turned over the task of preparing this risk estimate to a firm of engineering consultants contracted to prepare a *Preliminary Draft Environmental Impact Statement*. Their cancer risk assessment concluded that the maximum lifetime cancer risk from dioxin emitted by the Navy Yard incinerator would be 0.13 per million. Since EPA generally regards a risk of 1 per million acceptable, this result supported the DOS position that there is no significant risk from exposure to the incinerator's dioxin emissions.

The Williamsburg residents had no way of challenging this conclusion, derived as it was from an elaborate technical analysis hardly accessible to the public. At CBNS, however, we could critically review the DOS risk assessment—and it did not stand up under scrutiny. A major fault was the assumption that dioxin would enter the body only through the lungs, although an earlier study had shown that much more would be ingested, for example by children licking their dust-encrusted fingers. The risk assessment also ignored the fact that incinerators emit most of the 210 different compounds that fall under the common rubric of dioxin. Instead, the consultant estimated the risk from only the best-known compound, 2,3,7,8-TCDD.

Correcting for such defects, and accepting the DOS assumption about the amount of dioxin the incinerator would emit, we at CBNS concluded that the maximum lifetime cancer risk would be 29 per million rather than 0.13 per million—well above the 1-per-million guideline.

Pressed by the residents—and voters—of Williamsburg, the New York City Board of Estimate held public hearings at which these discrepancies and other objections to the incinerator were aired. The upshot was a Board of Estimate request that the DOS hire a new consultant to do another, "independent" analysis of the dioxin cancer risk. The new consultant adopted most of CBNS's corrections to the original risk assessment and reported that the maximum lifetime cancer risk would be 5.9 per million—a difference from CBNS's value of 29 per million readily encompassed by the uncertainties inherent in such estimates. In any case, the original DOS assessment now appeared to be between 45 and 223 times too low, and clearly in excess of the guideline.

One of the purposes of issuing an environmental impact statement as a "preliminary draft" is to allow for improvements before the final document is produced. In this respect, the procedure worked well, resulting in two separate reviews that corrected the original document's grossly underestimated assessment of the dioxin risk. This correction raised the issue of the incinerator's environmental acceptability, a matter that would presumably be addressed in further versions of the environmental impact statement. The DOS took an innovative, if astonishing, approach to this commendable process of amelioration: the dioxin risk assessment was simply eliminated from the subsequent draft, and eventually from the final environmental impact statement, perhaps on the view that no news is good news.

At this point, the incinerator project moved into the next administrative stage: extending hearings before an administrative law judge of the New York State Department of Environmental Conservation (DEC) in which the application for a construction permit was defended by the city and attacked by the opponents. The DEC noticed the strange omission of a cancer risk assessment in the final environmental impact statement and required the company that proposed to build the incinerator to produce one. Yet another consultant was hired. The new risk assessment concluded that the maximum lifetime cancer risk from the incinerator emissions would be 0.78 per million—once more conveniently within the guideline. This consultant took a very creative approach to the risk assessment. He decided that dioxin would take a novel path into peoples' bodies. Dioxin-contaminated fly ash would be emitted from the plant stack, then, descending to ground level, it would somehow mix with a ten-centimeter layer of soil. After being enormously diluted by the soil, the dioxin would come in contact with people through soil particles scattered into the environment.

There are a couple of things wrong with this picture. In the first place, dioxin will not readily penetrate the soil; it is so firmly bound to soil particles that, according to several studies, dioxin will hardly penetrate more than a few millimeters of soil when deposited on it as dust. It is also a fact that there is not much soil exposed to descending dust in

Brooklyn; houses and paving cover most of the borough. The absurdity of this concept suggests that it was invented in order to justify an enormous reduction in the computed exposure. A well-known feature of environmental impact statements is their considerable length, due to an enormous amount of often unnecessary, irrelevant discussion and computations. In this case, the consultant went to great trouble to determine how much dioxin people would absorb by eating the fish from the lake in Brooklyn's Prospect Park—something very few people are sufficiently rash to do. Correcting for its errors and absurdities, the new risk assessment actually leads to a maximum lifetime cancer risk of 12 per million—a rather good agreement with both the earlier estimates by CBNS and with the DOS's special dioxin report.

The debate over the Navy Yard incinerator's cancer risk assessment—a kind of technological tennis match with the DOS and its consultants on one side and the Williamsburg community (with considerable support from the New York Public Interest Research Group) and CBNS on the other—has helped to improve the method of making such assessments. Before the debate, there was little agreement over the relevance of a number of the crucial factors that influence the dioxin cancer risk, such as the route of exposure. Since then, apart from newly introduced notions, such as the invention of a bucolic, soil-covered Brooklyn, most incinerator cancer risk assessments have at least considered the relevant factors. Most of the results are similar, generally ranging upward from the 1-per-million guideline to 20 or more, indicating that according to this single criterion of health risk, the trash-burning incinerator is at best marginal and more likely unacceptable. The credit for this advance must go to the Williamsburg residents who have forced the municipality to test its hired consultants against independent analyses. The controversy has also encouraged new studies on the occurrence of dioxin in the environment. A recent one shows that dioxin is widespread in the air in Ohio, nearly all of it originating from incinerators that burn trash or sewage sludge.

The environmental controversies that have troubled the incinerator industry can be traced back to its power plant ancestry. Power plants are designed to convert as much as possible of their fuel's energy into steam and electricity. Hence their emphasis on high combustion efficiencies—that is, the completeness with which the fuel is burned and the energy released—that can range up to 99.999 percent. The industry engineers assumed that mixed trash could be burned as efficiently as coal or oil and that only 0.001 percent of its combustible material would be left unburned. They assumed that if the furnace temperature is high enough to combust it, any toxic material in the trash would also be 99.999 percent destroyed, eliminating any environmental hazard. This explains the claim that an incinerator operating at a high enough furnace temperature and combustion efficiency would destroy dioxin, and that those incinerators that emitted dioxin were poorly operated. Yet when CBNS examined the data from a series of incinerator tests, we found no statistically significant relation between dioxin emissions and either furnace temperature or combustion efficiency. Thus, the data failed to support the theory—derived from power plants—on which the incinerator engineers relied to control dioxin emissions. The incinerator engineers also failed to reckon with the extraordinary toxicity of dioxin, which renders exposure to soil contaminated with 0.0000001 percent of it (1 part per billion) unacceptable according to EPA guidelines.

Rather sobered by the fact that the industry was eagerly building huge incinerators without adequately understanding their operation, at least with respect to toxic emissions such as dioxin, at CBNS we went back to basics. In the chemical engineering literature, we found several reports of experiments indicating that fly ash can catalyze chemical reactions, including the addition of chlorine to unchlorinated dioxins. We also realized that the six-carbon rings (two of them) that make up dioxin and furan molecules occur in trash in the form of lignin, a common constituent of wood and therefore of paper. Lignin is likely to break down in the furnace, releasing these ring compounds, which, bound to fly ash particles, could then react with chlorine to form dioxins and furans. We also found a report of a Dutch investigator who showed that when paper was burned together with the chlorinated plastic polyvinyl chloride, dioxin was produced, whereas very little was produced when paper alone was burned. Finally, realizing that organic compounds such as lignin breakdown products would bind to fly ash only at relatively low temperatures (less than 500°C), we devised a hypothesis that dioxin, rather than being destroyed in the furnace, is actually *synthesized* in the cooler parts of the incinerator—that is, as the hot combustion gases flow from the furnace to the stack. So the crucial test of this theory was to measure the amount of dioxin in the flue gases leaving the furnace and later, when they had cooled down, at the base of the stack.

As it happened, the Canadian environmental agency, Environment Canada, which has a well-developed system of incinerator testing, produced the necessary data. In 1984, in an incinerator on Prince Edward Island, Canada, they found that while the dioxin content of the flue gas leaving the furnace was negligible, significant amounts were present in the gas entering the control device at the base of the stack. Dioxin must have been synthesized in between the furnace and the control device, in the cooler parts of the incinerator. It is now generally accepted, by the incinerator industry as well as government agencies, that dioxin is synthesized in trash-burning incinerators.

This conclusion transformed the problem of controlling the incinerator's environmental impact, at least with respect to dioxin. It meant that the incinerator *created* dioxin as it operated. Some dioxin is present in trash because paper is frequently contaminated with it as a result of chlorine bleach, which is often used to process wood pulp. Dioxin in such paper products is likely to be destroyed in the furnace if it is hot enough. But more is synthesized in the incinerator, and as a result, the trash-burning incinerator is a net producer of dioxin—an unintended dioxin factory. Depending on the efficiency of the control devices installed to precipitate fly ash, some of the newly formed dioxin will emerge from the stack into the air, while the remainder will be found in the fly ash trapped in the control device. In one form or another, the incinerator *creates* an environmental dioxin problem.

The industry has responded to this new understanding of how their incinerators work by introducing new control devices, generally a scrubber that cools and adds lime to the flue gases, followed by a fabric filter, which can usually capture 90 to 95 percent of the dioxin. While such controls correspondingly reduce the amount of dioxin emitted from the stack, they do so not by destroying it, but by transferring it to the fly ash captured by the scrubber and filter. Instead of entering the environment through the air, now most of the dioxin enters the environment when the fly ash is removed from the control device and disposed of.

Incinerator ash depositories vary a great deal. In Saugus, Massachusetts, incinerator ash (fly ash and the "bottom ash" that falls through the furnace grate) has been dumped on top of an old landfill, which itself is situated in a marsh. Residents are exposed to toxic materials as the ash is carried in the wind and its leachable components reach the water. In the town of Glen Cove, Long Island, incinerator ash was for a time piled in a parking lot, poorly protected from the elements. On the other hand, now most states' regulations require that, because of its toxicity, incinerator ash must be deposited in special landfills protected by liners and leachate control systems. But this adds considerably to the cost of operating the incinerator. Like the nuclear power plant before it, the trash-burning incinerator becomes progressively more costly as remedies are needed to control its inherent environmental defects.

A chief reason for the new regulations is the discovery that in addition to dioxin, fly ash is heavily contaminated with toxic metals, especially lead and cadmium. Lead occurs in discarded batteries, and to a lesser extent in electronic equipment. Cadmium also occurs in batteries and in certain plastics. These metals are released and vaporized in the heat of the incinerator furnace and then become trapped along with the fly ash in the control system. On the other hand, mercury, which also occurs in certain batteries, is so easily vaporized that most of it passes through the control system, out the incinerator stack, and into the air.

Thus, in the course of less than ten years a great deal has been learned about the environmental impact of trash-burning incinerators. Once regarded as a "proven technology" that created no environmental hazard, incinerators are now known to emit enough highly toxic compounds to create a risk of cancer and other diseases that is at best borderline, and more often unacceptable according to existing guidelines. The incinerator fly ash is so heavily contaminated with lead and cadmium as to frequently meet the EPA's official definition of a "hazardous substance" and therefore subject to very strict disposal rules.

Even if incinerator emissions do meet regulatory requirements, their impact on the environment is likely to exceed levels that have already been rejected in regard to other activities. For example, according to Dr. Peter Montague of the Environmental Research Foundation in Princeton, New Jersey, a proposed "state-of-the-art" incinerator for Falls Township, Pennsylvania, burning 2,250 tons of trash per day would emit 5 tons of lead annually. This equals the annual emissions from 2,500 automobiles using leaded gasoline. He points out that leaded gasoline is being phased out in part

part because of its health hazards, and asks, "Does it make sense to now burn garbage and introduce a new lead hazard?" Similarly, the proposed incinerator would emit 17 tons of mercury per year, an unwarranted environmental hazard when compared with the effort made by paper companies to reduce their emissions below 1 ton per year. Apart from these pollutants, the annual emissions from this incinerator would also include 580 pounds of cadmium, 580 pounds of nickel, 2,248 tons of nitrogen oxides, 853 tons of sulfur dioxide, 777 tons of hydrogen chloride, 87 tons of sulfuric acid, 18 tons of fluorides, and 98 tons of dust particles small enough to lodge permanently in the lungs. The incinerator's fly ash, which would contain a number of hazardous metals (lead, cadmium, chromium, and nickel), is to be deposited in a special landfill protected by a plastic liner to prevent leaching into groundwater. The metals will remain toxic for millennia; but the liner is guaranteed not to leak for only twenty years. Coincidentally perhaps, the legal responsibility of the incinerator operator, Wheelabrator Environmental Systems, for the ash also expires after twenty years.

Clearly, trash-burning incinerators have serious environmental defects. But they reveal a failing that is even worse: the incinerator industry has been building these devices without fully understanding how they operate, at least with respect to their impact on the environment.

In the absence of public opposition, the scientific challenges that led to this new knowledge would never have occurred. But the industry has hardly been grateful for this help. According to the industry and its collaborators, public opposition to incinerators is not an expression of a *social* concern but of a narrow, personal one: fear of the untoward effects on one's own health, one's immediate neighborhood, and the value of one's property. The incinerator industry's public relations experts have created a cute term—NIMBY (not in my backyard)—to convince us that the opposition to incinerators is simply a built-in, narrow-minded approach to any unpleasant intrusion into the neighborhood, a generic impulse to keep anything nasty out of one's backyard.

According to H. Lanier Hickman, Jr., the executive vice president of the Governmental Refuse Collection and Disposal Association, an organization that has been engaged in promoting incinerators, NIMBY is a social disease:

> The NIMBY syndrome is a public health problem of the first order. It is a recurring mental illness that continues to infect the public.

His answer to public opposition to incinerators is a "campaign to wipe out this disease."

To Calvin R. Brunner, a waste industry consultant, the real threat is anarchy:

> More than a century ago de Tocqueville warned us of may be [*sic*] too much democracy in America. Because everyone felt equal to everyone else, he projected that this would eventually lead to anarchy. . . . Is it possible that the NIMBY-ists will play a large part in proving de Tocqueville right in his assertions about democracy being an untenable form of government?

According to *Newsweek:*

> Forget Love Canal . . . Arrest the NIMBY patrols.

It turns out, however, that NIMBY is a myth. A detailed study of opposition to incinerators done for the State of California Waste Management Board tells us instead that:

> Public opposition to waste disposal facilities is a recent phenomenon. Prior to the rise of the environmental movement

> in the 1970's, waste facilities aroused little public concern, and rarely were facilities closed due to local opposition.

People have begun to worry about their backyards not because of a recent epidemic of antisocial selfishness, but—as the California report states—because "government and industry's failure to properly dispose of wastes received widespread publicity, which resulted in increasing public anxiety about the dangers associated with *all* waste facilities." What motivates the public in their opposition to incinerators is their concern not so much with the sanctity of their own backyard as with the quality of the environment that they share with the rest of society; this concern is not merely personal, but social as well.

In fact, the opposition to trash-burning incinerators has created a new arena for the exercise of democracy by generating debates about decisions ordinarily made with no public discussion at all. For example, the actual decision to build the Brooklyn Navy Yard incinerator was made by the Department of Sanitation, without public discussion, in 1979, three years before the first environmental information about the project was made public in the form of an environmental impact statement. In the absence of opposition from the Williamsburg community, that decision, even after the fact, would have gone unchallenged. Indeed, the heavy incinerator sales pressure has generated an equally intense community reaction. In almost every community, the decision to build an incinerator—or even to consider it—has been answered by the creation of an opposition group. The people who organize these groups are rarely environmentalists. They are, rather, the same people who will complain about a missing stop sign, a scarcity of police, or inadequate fire protection.

Community opposition has had an unmistakable impact on the incinerator industry. As early as 1984, the survey completed by the State of California Waste Management

Board concluded that "the most formidable obstacle to waste-to-energy facilities is public opposition." In April 1988, an investment firm that has been actively promoting bond issues that finance incinerators cautioned that

> public opposition will remain the toughest hurdle for this industry over the next several years. Concerns over dioxin emissions and ash residues high in heavy metal concentrations will continue to play a major role in this industry's development.

And a few months later the *Wall Street Journal* reported evidence of the public impact on the industry in terms that potential investors readily understand:

> More than $3 billion in projects have been scrapped in the last 18 months and new orders have slowed to a trickle.

This gloomy appraisal of the incinerator industry's outlook resulted from a series of victories chalked up by the community opposition groups. Thus, although the dispute over the Brooklyn Navy Yard remained unresolved in 1989, the mayoral election in that year has probably determined its fate. David Dinkins, who promised a moratorium on incinerator construction, defeated Edward Koch in the primary and Rudolph Giuliani in the general election—both avid proponents of incineration.

Similar battles have occurred in almost every community confronted with a proposed trash-burning incinerator; with increasing frequency, they have ended with the project killed. A noteworthy example is the LANCERS project—a proposal to build a 1,600-ton-per-day incinerator in Los Angeles. This incinerator was proposed in 1982 as the first of eleven projected plants in the Los Angeles Basin. The project was favored by the Los Angeles Bureau of Sanitation (BOS) because "the City needed a tried and true technology" for trash disposal. The project's Environmental Impact Report, issued in April 1985, offered evidence that the incinerator would meet this criterion; the BOS claimed that health effects would be so slight that weddings could be held on the plant's grassy lawn.

The residents of the neighborhood next to the plant site—a poor black and Hispanic community—were less enchanted with the project, despite an offered "community betterment fund" of $10 million that they would receive if it were built. They formed the Concerned Citizens of South Central Los Angeles to fight the incinerator. The main argument was over the plant's health effects, especially from dioxin in the air emissions and toxic metals in the ash. The Concerned Citizens mobilized help from Greenpeace, the Center for Law in the Public Interest, and from groups in other neighborhoods scheduled to receive incinerators. Most important, a group of students in the University of California Los Angeles Urban Planning Program, organized by two of their professors, Robert Gottlieb and Louis Blumberg, prepared a report on the plant's health risk assessment. It criticized the method used to evaluate dioxin toxicity, the inadequacy of the exposure analyses, and the failure to deal with the ash problem, and was widely publicized. In June 1987, after ten years and $12 million had been spent on the project, the mayor of Los Angeles, Tom Bradley, originally an ardent supporter of the project, decided to abandon it in favor of recycling. The basic reason was that voiced by a local member of Congress, who said that 95 percent of his constituents' letters opposed the LANCERS project.

While the details and the final outcomes vary, other battles over incinerators have followed the pattern common to the Brooklyn Navy Yard and LANCERS incinerators: a municipal decision to build an incinerator, well in advance of public discussion; community concern with its health effects; a public debate over this issue backed by technical support, on one side from the incinerator industry and on the other from the university and professional community; a final decision, made in political terms, based on the strongly voiced opinion of an electorate well informed by the debate. In Philadelphia, where an incinerator project strongly favored by the mayor was blocked by the City Council, technical help came from outside environmentalists brought in to council hearings. In San Diego, where the proposed incinerator was blocked by a referendum, a report on its health hazards by a group of local physicians played a crucial role.

Between 1985 and mid-1989, some 40 proposed incinerators were blocked, including projects in Los Angeles, Kansas City, Seattle, Boston, and Philadelphia. Some of the methods used to derail planned incinerators have been remarkably innovative. In San Diego, the proposed incinerator project was dropped when the local opposition group succeeded in passing a referendum that restricted the site of a proposed incinerator to a national monument, the city's famous zoo, or an area adjacent to the city's most expensive homes.

In June 1988, the *Wall Street Journal* noted the striking parallel between the demise of nuclear power and the difficulties encountered by the incinerator industry, and reported a sharp drop in incinerator orders between 1985 and 1987 due to public opposition. But in the last few years the parallel has broken down. Public opposition to nuclear power plants has succeeded in nearly shutting down this industry, but has failed to put any less hazardous power-producing technology in its place. In contrast, the public campaigns against incinerators have gone beyond simple opposition to vigorous efforts in favor of a far better means of trash disposal—recycling.

Recycling cures the ecologically harmful linearity of modern production technologies. Glass provides a good example. Conventionally, a linear series of processes produces a glass object, let us say a pickle jar—and converts it into trash. At the start, the glass factory heats sand, lime, and some minor ingredients, converting them into molten glass, which is then fashioned into various objects such as the containers that represent about 90 percent of the glass found in trash. The containers are sold to other producers (chiefly of food products) that may use them to transport beer, mayonnaise, jam, or pickles to their ultimate consumers. Here the container is separated from its contents and, of no further use to the householder, is discarded into the kitchen trash pail. There it joins food scraps, soiled paper towels, tin and aluminum cans, plastic wrap and aluminum foil, yesterday's newspaper, discarded junk mail and packaging, and a host of other items that constitute the melange of trash. Finally, the trash—usually encased in a plastic bag—is taken to the curb for collection, then carted to a landfill or incinerator—depositories that, as we have seen, create serious environmental hazards.

Now let us revise this melancholy scenario. Suppose we capture, in a freeze-frame, the moment in which the glass jar is emptied of its last pickle. At this brief moment in its history, the container is still a useful object, for it is, after all, the same container that the pickle factory bought from the glass company. The container's potential usefulness could be realized by returning it to the pickle factory—as was once done with returnable beer bottles. Alternatively, the empty container could be returned to the glass manufacturer, where it would be remelted and formed into some other product or even into an identical jar that could once more carry pickles to their destiny. In either case, the once linear process is converted into a circular one. This eliminates the detrimental environmental impact that occurs at the end of the line, for it prevents the conversion of the otherwise useful container into trash.

In sum, at the moment it is emptied, poised in the householder's hand, the pickle jar is at the brink of two alternative fates. If it is tossed into the trash pail, the jar is transformed into trash, fated to contribute to all the ensuing environmental problems. On the other hand, if—perhaps first rinsed—the jar begins a voyage back to the pickle company or the glass factory, its fate is environmentally benign.

Indeed, it is even beneficial. Glass manufacturing itself has an impact on the environment, for example from the pollutants generated by the fuel burned to melt the starting ingredients. About 25 percent less fuel is needed to produce molten glass from crushed glass than from sand, so that to the extent that the glass factory uses recycled glass, it burns less fuel and creates that much less pollution. Thus, recycling glass instead of discarding it produces two kinds of environmental benefit: it prevents the production of trash and thereby reduces the environmental hazards incident to trash disposal, and it reduces the environmental impact incurred by producing glass from virgin raw materials.

The feasibility of such recycling schemes is, of course, affected by economic considerations. When breweries were local, returnable bottles were preferred because they made about forty return trips before replacements were needed. This cut down on bottle purchases, a saving greater than the cost of hauling back the empty bottles and washing them. With not only beer but mayonnaise, jams, and pickles produced by highly centralized companies, distant from consumers, economics often dictates against recycling by reuse. On the other hand, once the pickle jar is recycled and converted into crushed glass, it becomes more broadly useful to glass companies, which will always save money—for example, by saving fuel—by using it in place of manufacturing virgin glass. In theory, then, glass containers could be infinitely recycled, manufactured entirely from the glass contained in recovered containers. In practice, U.S. glass container companies now generally add about 25 percent crushed, recycled glass to the virgin raw materials. But they could readily use more; 53 percent recycling of glass has been attained in Holland. If a proper recycling system were established, glass containers could be made exclusively from old ones.

Recycling of other trash constituents is more complicated, because unlike glass, which is chemically unchanged by its uses, other materials are modified when they are used to manufacture trash components and some additional steps must be taken to recover the material in its original form. For example, when steel is used to make "tin" cans, it is plated with tin. If the recycled can is simply melted at the steel plant, its further use is hindered by the admixture of tin in the molten steel. Consequently, cans must be detinned before they become usable scrap steel, which is then readily melted and manufactured into the numerous objects that are made of steel—including tin cans. The process is still economically sound, because even after the cost of detinning (which is reduced by the sale of the recovered tin) is subtracted from the value of the remaining metal, steel made from that metal is still cheaper than virgin steel made from iron ore. This recycling process is also ecologically sound, for the avoided environmental impact—the heavy pollution engendered by mining iron ore and manufacturing steel from it—is greater than the impact of detinning.

Recycling paper encounters an additional complexity. When discarded paper is converted into pulp in the process of being manufactured into new paper, some of its constituent fibers are broken. This reduces the quality of the paper that can be made from it. Consequently, unlike glass, steel, or aluminum, paper recovered from trash often cannot be recycled indefinitely. For example, progressively recycled, newsprint will lose some of its strength and may not behave well in high-speed presses, a difficulty that can be minimized by mixing recycled newsprint with virgin pulp. The *Los Angeles Times* is printed on paper that contains about 80 percent recycled material. Several states plan to require local newspapers to use some proportion of recycled fibers in their paper. But the American Newspaper Publishers Association has objected to such legislation as an attack on the freedom of the press. Paper recovered from trash can also be converted into less exacting products, for example toilet tissue, in which fiber length is not so important. In paper recycling, this results in a kind of cascading of the recovered products downward in quality—and in value. Nevertheless, it is still cost-effective for paper plants to use paper recovered from trash, at least in part, to manufacture their products. This is especially true if different qualities of the trashed paper can be kept apart, in the form of separated newspapers, computer paper, or cardboard.

Most people are surprised to learn that food garbage is also recyclable. Unlike the other recyclable components of trash—glass containers, metal cans, or paper—food garbage must be transformed in its chemical composition when it is recycled. In a fundamental sense, food is derived from the soil. In nature the plants and animals that comprise food return to the soil when they die, where soil microorganisms decompose them and transform their component substances into an important soil constituent: humus. These microorganisms and the processes they mediate are readily domesticated. This occurs, for example, in the compost piles that farmers and gardeners have long used to convert plant and animal residues, including manure, into a useful soil additive of significant nutritive value.

Public and professional interest in recycling has varied considerably over the years. Dealing in junk—more politely, "secondary materials"—is an old and continuing form of commerce. But except for recovering the occasional appliance or bedspring discarded by households, it has had little impact on residential trash disposal, concentrating instead on scrap metals and defunct cars. During World War II, when supplies of critical materials were restricted, with government encouragement Americans adopted recycling as a normal thing to do. Old pots and pans and flattened tin cans were set out for collection; newspapers and cardboard were recycled; rubber bands were assembled into huge balls. But this was a wartime enthusiasm, and from the end of the war until the ecological outburst of Earth Day 1970, there was no large-scale recycling.

In the ecological spring of 1970 and succeeding years, many people looked for some way to contribute personally to environmental quality. Recycling was a ready answer. Ecology-minded community groups, scout troops, and churches organized depots where people were encouraged to bring their discarded cans, bottles, and newspapers. Apart from aluminum cans—which were so valuable to aluminum companies that they took on the task of collecting them—most of these efforts sooner or later collapsed, frustrated by the vagaries of the secondary materials market.

Items collected at the recycling centers were usually sold to brokers for resale to the actual users. Whenever commodity prices fell—for example, in the recession years of the early 1980s—the value of secondary materials frequently fell to zero. Brokers, who like all private entrepreneurs must operate at a profit, would then stop buying the collected cans and paper. The recycling depot would clog up and be forced to close. This experience has encumbered recycling with a "market problem" that is frequently cited by incinerator manufacturers to prove that the process is not a realistic means of trash disposal.

In the last few years, with the advent of a publicly recognized "trash crisis," and the lively discussions accompanying the widespread opposition to incineration, interest in recycling has suddenly revived. The revival is almost entirely due to the same groups that have opposed incinerators. Where they have succeeded in stopping an incinerator project, they have accepted the responsibility of proposing recycling as an alternative—confounding the incinerator industry's claims that they are "NIMBYists" opposed to any unwanted intrusion.

A major issue is whether a recycling program could possibly dispose of as much trash as an incinerator (which generally burns about 70 percent by weight and leaves a 30 percent residue of ash) and could therefore serve as an alternate trash disposal technology. A survey of seventeen curbside recycling programs in 1989 by the National Solid Wastes Management Association reported that they recycled an average of only 15 percent of the residential trash stream. Yet about 90 percent of the trash components are *capable* of being recycled by existing methods. But that was a theoretical figure that had never been approximated in practice, at least in the United States.

At CBNS, converting this theoretical figure into practice was a welcome challenge. The key appeared to be what was done with food scraps—garbage. If such sticky stuff is mixed with recyclable components like cans and bottles, it becomes difficult to get those containers clean enough to produce good-quality metal or glass. Moreover, if other trash components are mixed with the food garbage, the compost made out of it is likely to contain bits of glass, plastic, and metal—not a good situation for the home gardener who might want to use it. So when the town of East Hampton, Long Island, asked for advice about trash disposal, we designed a household separation system based on four containers: one for food garbage and soiled paper, which together with yard waste can be readily composted; a second for all forms of paper and cardboard; a third for cans and bottles; and a fourth for the nonrecyclable remainder of the trash. The first three containers held all the recyclable components. In recent years, "materials recovery facilities" have been developed, which, using mechanical devices and hand separation, can further separate the nonfood recyclables collected from households into separate grades of paper, aluminum cans, tin cans, and color-sorted crushed glass (clear, green, and amber). Similarly, fairly simple compost facilities can convert separated food garbage and soiled paper (together with yard waste) into useful compost. Thus, a scheme based on the four-container household separation, followed by processing at these facilities, seemed theoretically capable of recovering most of the recyclable trash. But that would depend on at least two unknown factors: how well householders would do the job of separation into the four containers, and the efficiency with which the processing facilities would convert the collected recyclables into marketable products.

With the help of the town, a practical test was set up with one hundred volunteer households. They were provided with sets of containers and instruction sheets. A yellow-and-white-striped party tent was erected next to the town landfill—where residents normally took their trash—and for ten weeks containers were brought in, weighed, and distributed to large bins. At the end of the test, it turned out that the nonrecyclables amounted to less than 13 percent of the total trash, confirming the expectation that nearly 90 percent of the trash is indeed recyclable. Several different methods were used to compost the food garbage, together with yard waste and sludge from the town sewage treatment plant. In the simplest method, the food garbage was ground, mixed with grass clippings, wood chips, and sludge. The pile was mixed from time to time with a mechanical scoop in order to aerate it sufficiently to support the necessary microbial activity. After several months, the compost was screened, yielding a product that, when analyzed for toxic metals and nutrients, passed muster as a marketable product, suitable for sale to gardeners and horticulturists.

The other recyclables were taken to a cooperating materials recovery facility in Groton, Connecticut. There the East Hampton material, after first being weighed, was sorted. Then, the material rejected during processing having been weighed as well (a horseshoe turned up in the paper), it was possible to calculate the overall physical efficiency of the separation and sorting process. It turned out that 84.4 percent of the trash originally collected from the participating households had been recovered, in the form of three grades of paper, aluminum cans, tin cans, color-sorted glass, scrap metal, and compost. In addition, hardy CBNS staff members sorted through representative batches of the nonrecyclables and found that certain recyclables had been mistakenly placed in container 4. Since this amounted to 2.4 percent of the total trash stream, the overall data showed that 86.8 percent of the trash consisted of recyclable materials and that the householders and the processing facility had done their job of properly sorting them with an admirable efficiency of 97 percent (84.4 percent divided by 86.8 percent).

Based on these results, a full-scale intensive recycling system was designed for the town. Assuming that even with a town ordinance mandating recycling, 10 percent of the year-round residents and commercial establishments would manage to avoid this responsibility and that the numerous summer residents would be somewhat less cooperative, we estimated that the system would recycle about 70 percent of the total trash—a figure that matched an incinerator's disposal capacity. Moreover, the recycling system would cost about 35 percent less than the cost of building, financing, and operating an incinerator. Like all other Long Island towns, East Hampton was required by New York State to file a trash disposal plan that would sharply reduce landfill requirements. In April 1989, the Town Board voted to adopt the intensive recycling system as its means of trash disposal, and has now begun the process of financing and building the system.

The failure of earlier programs to overcome the vagaries of the market for secondary materials is often cited as an argument against recycling. It is pointed out, for example, that many of the earlier programs were forced to close because they no longer made a profit when the price of secondary materials fell, often to zero. However, a trash disposal system is recognized as a *cost* to the community and is certainly not expected to run at a profit. An intensive recycling system such as the one developed for East Hampton is still less costly than incineration, even if all of its recycled materials were simply given away. In fact, it would still be cost-effective if—as happened on Long Island in 1989—the recycled newspaper market was so glutted that brokers *charged* $20 per ton to accept them. Thus, because it is a total system of trash disposal, intensive recycling can overcome difficulties inherent in the marketing of secondary materials.

The market for compost made from food scraps and yard waste is essentially unlimited. Compost is a form of humus, an important organic ingredient of virgin soil. The United States loses topsoil through erosion at the astonishing rate of about 3 billion tons annually, a process that obviously cannot be allowed to go on indefinitely. Compost properly prepared from separated trash would be free of toxic materials and suitable as an additive to agricultural soils. Nationally, about 40 million tons could be produced annually and could readily find a market as a means of combatting erosion, albeit at a cost for the necessary transportation.

Another argument against recycling is that in order to be effective, public participation in the initial household separation must be high—at least 80 to 90 percent. Small towns such as Hamburg, New York, or Woodbury, New Jersey, have already achieved this level of participation. According to a 1989 survey, the average level of participation in some fifteen communities with active (although partial) recycling programs is 73 percent. These levels rise as the public becomes more educated about the advantages of recycling. In Seattle the level of participation is currently 63 percent, and 34 percent of the trash stream is being recycled.

It is sometimes suggested that a high rate of participation will be impossible in a large city because of the considerable "cultural" differences among the residents. In the course of a CBNS recycling study in Buffalo, New York (population 320,000), there was an opportunity to analyze this problem. The city had established a voluntary pilot recycling program in three demographically different neighborhoods, and with the collaboration of Citizen Action of Buffalo (a local community group), a CBNS team determined the rate of participation in each of them. The overall rate of participation was 63 percent, a figure close to the average of 67 percent for a series of communities ranging in population from 6,000 to 180,000. However, the rate in one of the neighborhoods, Delaware, was 86 percent, a rate statistically different from those of the other two neighborhoods, Fillmore (54 percent) and Masten (45 percent). According to census data, Delaware is 94 percent white; the median household income is $15,835; 73 percent of the residents are high school graduates. Fillmore is 83 percent white and 15 percent black; the median household income is $10,184; 37 percent of the residents are high school graduates. Masten is 94 percent black; the median household income is $9,511; 47 percent are high school graduates. It appears that the statistically significant difference in participation rate between Delaware and the other two neighborhoods may reflect a comparable difference in family income. On the other hand, there is no evident relationship between participation rate and education and race. Fillmore and Masten have nearly equal participation rates (they are not statistically different), yet differ considerably in racial composition. Although Delaware residents have a higher level of education than the residents of Fillmore and Masten, in Masten there is a higher percentage of high school graduates and twice the percentage of residents with one to three years of college than in Fillmore. Yet Masten and Fillmore have essentially the same rates of participation.

The Buffalo study also showed that the city's economy would benefit from recycling, because most of the expenditures—for example, for the operation of compost and materials recovery facilities—would create jobs and contribute to the city's economy. In contrast, fees paid to an incinerator operator, generally a large outside corporation, would leave the city and not add to its economy.

It might be noted that the Buffalo recycling program has not as yet developed a strong educational component, which experience shows is essential to achieve good participation rates. Detailed information about recycling procedures could overcome difficulties that may be associated with low income—for example, a kitchen too small to readily accommodate more than one trash receptacle, unless they are designed to stack vertically.

Proponents sometimes argue that incineration is an available, "proven" technology, whereas the ability of recycling to serve as a means of trash disposal remains to be proven. The only thing "new" about intensive recycling is that it is consciously intended to recover all the recyclable components of trash. Its component parts are existing, well-established procedures. Household separation of trash into three or more containers was once the *mandated* procedure in a city as large as Los Angeles (which appears to be on the way to reestablishing it), and is an everyday practice in dozens of American cities such as Seattle. Composting is an ancient "low-tech" process, readily practiced in backyard piles or, as in a number of European communities, on a large scale with mechanical equipment. A dozen or more materials recovery facilities have been operating over the last decade to recover paper, metals, and glass from community recycling systems. The importance of the intensive recycling demonstration in East Hampton lay in a simple innovation: it brought these existing procedures to bear on the task of disposing of *all* of the trash recyclables.

It would appear, therefore, that there is now a choice of trash disposal technologies that can sharply reduce reliance on landfills: incineration or *intensive* recycling. These alternatives reflect the basic division between the competing strategies of dealing with pollution: reliance on controls to mitigate the impact of pollutants, or preventing their production to begin with. The moment at which the householder has emptied the pickle jar marks the dividing point. If the jar is tossed into a common bin of household discards, it *becomes* trash, an environmental pollutant that must be controlled to diminish its ecological impact. Consigning the jar instead to its proper recycling container retains its value as glass and prevents its conversion into trash.

Conventional recycling systems cannot compete with incineration because they are partial—aiming to deal with only part of the recyclable components of trash (generally newspapers, cans, and bottles). This limits the amount of the trash stream that can be disposed of, so that incineration is necessary as well. To compete with incineration, recycling must be intensive—that is, targeted on *all* the recyclable components. Both intensive recycling and incineration leave residues that must be otherwise disposed of, generally to a landfill. However, the landfilled residue from intensive recycling is inert material, chiefly plastic. The residue from incineration is toxic ash, which is difficult to landfill or dispose of in any other way without creating new environmental problems.

Thus, the choice between incineration and intensive recycling now confronts the state and federal agencies of environmental protection, and the municipalities and towns that they serve. Many of them have managed to evade this decision, however. A common response to the growing evidence of the merits of recycling has been "integrated waste management." This has a nice ring to it, proposing that trash disposal should be accomplished by joining recycling, incineration, and landfilling into a single system. In New York State, for example, communities are told that the preferred methods of disposal are waste reduction and recycling, with incineration in third place and landfilling least desirable.

However, the state nevertheless urges communities to build incinerators, providing that they have filed a plan for recycling as well; a plan that calls for recycling 15 to 25 percent of the trash is regarded as satisfactory. The excuse for favoring incineration, despite its low position in the order of preference, is that it is necessary as a transitional phase, before recycling is sufficiently developed.

In fact, the only insurmountable hindrance to recycling is building an incinerator. For a simple physical reason, incineration actually *interferes* with recycling: about 80 percent of the trash stream consists of components, such as paper and food garbage, that can either be burned or recycled, but obviously not both. Modern incinerators are expected to last for twenty to thirty years and must be fueled with trash to 85 percent of capacity so that sufficient steam or electricity can be sold to allow them to operate in the black. In fact, most incinerator contracts require the town to supply the incinerator with enough trash to permit economic operation—or failing that, to pay for the lost revenue from energy sales. In New Jersey, the state's Department of Environmental Protection has ignored the clash between incineration and recycling, and has urged counties to build incinerators. As a result, a Warren County incinerator has been forced to operate, uneconomically, well below capacity because burnable trash components have been diverted to conform to a state law requiring 25 percent recycling. The proposed solution is to reduce the size of future incinerators, but that would freeze recycling at 25 percent—again, uneconomically—for the twenty-year lifetime of the incinerators.

By including recycling, "integrated waste management" takes on a kind of ecological allure, even acknowledging that it is the *preferred* way to deal with trash. But it is important to identify the problem that recycling answers. If the problem is to give people a sense of ecological virtue, then any token amount of recycling—for example, the 25 percent required by law in New Jersey, New York City, and elsewhere—will do. If the problem is trash disposal, then recycling must be intensive and designed to deal with the 90 percent of trash that is recyclable—a situation rendered impossible by any combination with incinerators.

The trash crisis illustrates the melancholy environmental picture—and shows what can be done to brighten it. To begin with, it is a prime example of the general failure, thus far, to significantly improve environmental quality. Of the present disposal methods—about 80 percent to landfills, about 10 percent incinerated, and the rest recycled—landfilling and incineration not only fail to deal with this problem but make it worse. Second, like pollution generally, most of the enormous stream of trash, especially its most rapidly increasing segment—plastic and other packaging—clearly originates in the system of production: in the design of the brewing and bottling industries; in the rapid growth of the plastics industry; in the increased use of battery-driven appliances and toys. Third, the disposal method favored by industry, by environmental agencies, and, until recently, by most municipal officials—incineration—is an effort to *control* the environmental impact of the pollutant, trash, that, like the control strategy generally, generates more environmental problems than it solves. Moreover, the effort to cope with the incinerator's serious environmental faults has (as we have seen in chapter 4) provoked regulatory efforts, such as the "linguistic detoxification" of ash and palpably specious attempts to downgrade the toxicity of dioxin, that have seriously eroded the integrity of federal and state environmental programs.

In contrast, intensive recycling, a method of trash disposal that relies on prevention rather than control—like the other, still scarce examples of prevention—is clearly far superior to incineration in its environmental impact. In addition, like the other examples of ecologically sound technologies cited in chapter 5, intensive recycling is also more cost-effective than incineration and, in fact, stimulates rather than drains a community's economy. Finally, the recent history of the trash crisis, in particular the growing movement away from incineration toward recycling, is an encouraging sign that despite the powerful taboo against it, it is possible for the public to intervene in the hitherto private decisions that determine production technologies. Of course, these efforts have been aided by the fact that official agencies, which unlike private corporations are open to public influence, are involved in trash-disposal decisions. Nevertheless, the self-organized groups of citizens who, with increasing frequency, have blocked proposed incinerators and have moved their communities toward recycling, have demonstrated the power of environmental democracy. They are the pioneers in the historic transformation that alone can restore the quality of the environment.

7

POPULATION AND POVERTY

ONE OF THE VIRTUES of the environmental point of view is that we see the planet as a harmonious whole, a global system of water, soil, and living things bounded by the thin skin of air. However, when we look at the planet with an eye on human manifestations—the technosphere and the social systems that create it—it is split in two. The northern hemisphere contains most of the modern technosphere—its factories, power plants, automotive vehicles, and petrochemical plants—and the wealth that it generates. The southern hemisphere contains most of the people, nearly all of them desperately poor.

The result of this division is a painful global irony: the poor countries of the south, while deprived of an equitable share of the world's wealth, suffer the environmental hazards generated by the creation of that wealth in the north. The developing countries of the south will not only experience the impact of global warming and ozone depletion, which are now chiefly due to the industrialized countries, but are also victimized by the north's toxic exports. For example, as bans have been imposed on particularly dangerous pesticides in industrialized countries, manufacturers have marketed them in developing countries instead. There, poorly regulated, they have created in the bodies of local populations the world's highest concentrations of pesticides. Similarly, as environmental concerns have limited disposal sites for trash and the toxic ash from trash-burning incinerators in the United States, efforts have been made to get rid of these pollutants—not always successfully—in developing countries.

Yet the gravest threat of the environmental crisis to developing countries comes, not from the pollutants so generously imposed on them by their wealthy planetary neighbors, but from a more subtle source. This threat arises from a serious, frequently voiced misconception about the origin of the environmental crisis. In this and earlier analyses, I have argued that the environmental crisis originates, not in the natural ecosphere, but in the man-made technosphere. The data about both the development of the post-1950 assault on the environment and the effort since 1970 to reduce it support this conclusion. There is, however, another view of the environmental crisis that turns these relationships upside down. This view holds that the problem is ecological; that environmental degradation originates in an imbalance between the earth's limited resources and the rapidly growing human population, which stresses the environment and also causes social problems such as poverty and hunger.

This position had a popular following in the early days of the environmental movement, based on unequivocal assertions by some well-known environmentalists. In a widely quoted article, "The Tragedy of the Commons," Garrett Hardin put it this way:

> The pollution problem is a consequence of population. It did not matter much how a lonely American frontiersman disposed of his waste. . . . But as population became denser, the natural chemical and biological recycling processes became overloaded. . . . Freedom to breed will bring ruin to all.

Paul Ehrlich's best-seller, *The Population Bomb,* was even more explicit about the origin of the environmental crisis:

> The causal chain of the deterioration [of the environment] is easily followed to its source. Too many cars, too many factories, too much detergent, too much pesticide, multiplying contrails, inadequate sewage treatment plants, too little water, too much carbon dioxide—all can be traced easily to *too many people.*

In the early 1970s, such statements encouraged the view that population control is the only practical way to reduce pollution, a notion that some environmentalists took personally. Visiting a midwestern university in the early 1970s, I recall a conversation with a pregnant faculty member who had received anonymous letters condemning her for contributing to the environmental crisis. Enthusiasm for contraception as an environmental strategy has faded considerably, although recently the head of the National Organization for Women (NOW), desperate for allies in the fight against the Supreme Court's invitation to state control of abortions, sought to enlist environmentalists on these same grounds. This basic position is still held by a number of leaders of the environmental organizations that have grown to prominence since the early 1970s. Russell W. Peterson, the former president of the National Audubon Society, a major environmental organization, expressed it this way a few years ago:

> Almost every environmental problem, almost every social and political problem as well, either stems from or is exacerbated by the growth of human population. . . . As any wildlife biologist knows, once a species reproduces itself beyond the carrying capacity of its habitat, natural checks and balances come into play. . . . The human species is governed by this same natural law. And there are signs in many parts of the world today—Ethiopia is only one of many places, a tip of the iceberg—that we *Homo sapiens* are beginning to exceed the carrying capacity of the planet.

Such statements send a chilling message to developing countries. They are in a desperate struggle to improve living standards, and—in violation of Mr. Peterson's dictum—are eager to use *more* resources to support their rapidly growing populations. Surely sweeping prescriptions such as those just cited, which affect the destiny of most of the world's people, ought to be solidly founded in fact. Are they?

The chief source of these views is a fundamental ecological concept regarding the relationship between the eater and the eaten. In a normal ecological system, there is a balance between a species' population and its food supply. If rabbits reproduce at the rate at which they are eaten by wolves and the wolves reproduce at the rate at which they die or are killed by hunters, both populations will be stable in size. Suppose, however, that some outside influence on the wolves' death rate is eased—fewer hunters, perhaps—so that their population rises above the equilibrium size. Now the rabbits are likely to be eaten faster than they can reproduce, and their population will decline, reducing the "carrying capacity" of the ecosystem for wolves. The wolf population will then be short of food, die off faster, and become smaller until it is once more in balance with the rabbits.

Applied to human beings, this concept suggests that in a country like Ethiopia, devastated by repeated famines, starvation is a symptom of overpopulation, and—in the absence of outside intervention—a precursor to a catastrophic population decline that will restore the balance between its size and the country's limited food supply. Relying on this concept, some environmentalists urge population reduction and oppose famine relief as a misguided, futile gesture. Thus, Garrett Hardin provides this explanation of why he is opposed to feeding hungry countries:

> When we send food to a starving population that has already grown beyond the environment's carrying capacity we become a partner in the devastation of their land. Food from the outside keeps more natives alive; these demand more food and fuel; greater demand causes the community to transgress the carrying capacity more, and transgression results in lowering the carrying capacity in the future. The deficit grows exponentially. Gifts of food to an overpopulated country boomerang, increasing starvation over the long run. Our choice is really between letting some die this year and letting more die in the following years. . . . Only one thing can really help a poor country: population control.

Apart from humane considerations, there is a ready response to this position: if the necessary funds were available, by applying modern production processes, Ethiopia could increase food production and if need be use its increased wealth to import it. But such reliance on economic growth and development is also regarded as ecologically unsound by population-minded environmentalists, for in Paul Ehrlich's terms, it would only lead to "too many cars, too many factories, too much detergent, too much pesticide" and the inevitable deterioration of the environment.

This approach has been elaborated on a global scale by means of computer models that purport to show, mathematically, that the interaction between a growing world population, the economic growth impelled by it, and the resultant environmental degradation and food shortages leads inevitably to the kind of population crash experienced by wolves in an ecosystem of too few rabbits. This conclusion was reached by the Club of Rome (a self-appointed organization of industrialists and environmentalists) in a report that was widely publicized, but less widely acclaimed for its scientific soundness, *The Limits to Growth.*

The "limits to growth" approach is based on a serious misconception about the global ecosystem. It depends upon the idea that the Earth is like a spaceship, a closed system isolated from all outside sources of support and necessarily sustained only by its own limited resources. But the ecosphere is not in fact a closed, isolated system, for it is totally dependent on the huge influx of energy from an outside source—the sun. Living things must be provided with energy to sustain their vital processes, in particular growth, development, and reproduction. That energy is derived from the sun. Sunlight, absorbed by plants, drives the energy-requiring chemical reactions that synthesize the complex organic compounds characteristic of life, such as protein, carbohydrates, fat, and nucleic acids. This process, photosynthesis, is the gateway that brings solar energy into the ecosphere: the rabbit, nibbling vegetation, derives its energy from the plant's organic compounds; the wolf obtains its energy by devouring the rabbit. Such food chains transmit the solar energy initially captured by plants throughout the ecosphere.

Thus, solar energy, captured by photosynthesis, sustains every form of life and drives the ecological cycles in which they participate. If an ecological cycle is viewed only as a static array of animals, plants, and microorganisms linked through the physical environment into a circular system, it appears to be closed, like a ring. But this image is misleading, for without the energy that it receives, externally, from the sun, the plants and animals would die and the circular system would disintegrate.

Solar energy also creates the weather: the seasonal temperature changes; the moisture that the sun lifts from the oceans; the wind, and the storms that carry rain and snow to the soil and replenish the lakes and rivers that feed the oceans. In turn, the weather molds the physical features of the Earth's surface, creating the ecological niches that living things occupy. In sum, the global ecosystem is not, in the basic thermodynamic sense, an isolated, self-sufficient system. In fact, neither is a spaceship, which after all depends for the energy that operates it on electricity generated by photovoltaic cells—from the sun.

In the abstract sense, there is a global "limit to growth," but this is determined not by the present availability of resources but by a distant limit to the availability of solar energy. It is true, of course, that the ecosystem that occupies the Earth's thin skin, and the underlying mineral deposits, are essential both to population growth and economic production. It is also true that there is a potential limit to economic growth due to the finite amounts of these essential resources. However, since matter is, after all, indestructible, the chemical elements that comprise the planet's resources can be recycled and reused indefinitely, as long as the energy necessary to collect and refine them is available. This is precisely what is done when the resource is sufficiently valuable; despite its extensive dispersion, well over half of all the gold ever mined is still in hand today, regathered when necessary by expending energy. Hence, the ultimate limit on economic growth is imposed by the rate at which renewable, solar energy can be captured and used. If we ignore the exceedingly slow extinction of the sun, this limit is governed only by the finite surface of the Earth, which determines how much of the energy radiated by the sun is actually intercepted and is therefore capable of being used.

Thus, the theoretical limit to the growth of the global economy is determined by the rate at which the Earth receives solar energy. How close is this limit at present? It has been estimated that the solar energy that falls annually only on the Earth's *land* surface is more than a thousand times the amount of energy (almost entirely from fuels, hydroelectric and nuclear power) now being used each year to support the global economy. Of course, because some parts of the land are difficult to reach or otherwise unsuitable, not all of the solar energy that falls on it could be used. If, let us say, only 10 percent of the total solar energy falling on land could be captured, it would still be possible to expand our present rate of using energy a hundredfold before encountering the theoretical limit to growth. Even if this figure should turn out to be somewhat optimistic, it seems clear that we are at present nowhere near the limit that the availability of solar energy will eventually impose on production and economic growth. That distant limit is irrelevant to current policy.

The issue we face, then, is not how to facilitate environmental quality by limiting economic development and population growth, but how to create a system of production that can grow and develop in harmony with the environment. The question is whether we can produce bountiful harvests, productive machinery, rapid transportation, and decent human dwellings sufficient to support the world population without despoiling the environment.

It is useful, at this point, to turn to the data that relate environmental deterioration to the factors that influence it. As we have seen, production technologies differ considerably in their tendency to pollute the environment. Consider, for example, the different environmental impacts of two alternative technologies of beer distribution: the throwaway bottles, and the returnable bottles that are likely to be used forty times before being broken or discarded. Each bottle contains an economic good—twelve ounces of beer—and the production of that good is associated with a pollutant: the bottle discarded as trash. We can compare the pollution-generating potential of the two technologies by computing the number of beer bottles used to deliver each twelve ounces of beer. In the case of the throwaway bottles, the figure is one bottle per twelve ounces of beer; for returnable bottles, the figure is 1⁄40 bottle per twelve ounces of beer. Thus, the pollution-generating tendency of a technology can be expressed numerically as the amount of pollution generated in producing a unit amount of economic good.

The total amount of pollution generated can then be expressed by multiplying this "technology factor" (pollution per unit good) by the total amount of good produced. Finally, the latter figure can be broken down into the product of two factors: good produced per capita (the "affluence factor") multiplied by the size of the population. In this way, the total amount of pollution can be expressed numerically in the form of an equation:

total pollution = pollution per unit good ×
good per capita × population

This relationship shows that the total amount of pollution generated will increase if any of the three factors increases. Thus, it is possible to say with equal validity that environmental deterioration—say, the number of beer bottles—is exacerbated by "too many people" as Paul Ehrlich claims, or, in keeping with my own analysis, by a change in the technology of production that increases the number of bottles used to deliver twelve ounces of beer. And it is equally possible that the number of discarded beer bottles will increase because of greater beer consumption per capita. What is at issue is the relative impact of each of these factors on environmental pollution. Such an evaluation will indicate which factor, if reduced, provides the most effective means of improving environmental quality.

In the United States, data are readily available to evaluate the relative effect of the three factors on a number of pollutants. In the case of beer bottles, they show, for example, that between 1950 and 1967, as the number produced annually increased by 593 percent, the population increased by 30 percent, per capita beer consumption (the affluence factor) rose by 5 percent, and the number of bottles used per unit of beer shipped (the technology factor) increased by 408 percent. Clearly, the largest impact on the amount of beer bottle trash was due to the technology factor: the introduction of nonreturnable bottles, which sharply increased the number of bottles needed to ship a unit amount of beer.

The relative impact of the three factors in reducing the environmental impact of beer bottle trash between 1950 and 1967 is indicated by the following: if there had been no change in the technology of beer distribution, the number of beer bottles would have increased by 37 percent; if only population had remained constant, the number of beer bottles would have increased by 433 percent; if only the affluence factor had remained constant, the number of beer bottles would have increased by 560 percent. Clearly, an effort to change the technology of beer distribution will result in the greatest reduction in environmental impact.

The pattern revealed by the beer bottle data is typical of the new post-1950 production technologies. For example, between 1950 and 1967, when pesticide use for crop production increased by 266 percent, the population increased by 30 percent, crop production per capita (the affluence factor) by 5 percent, and the amount of pesticide used per unit of crop production (the technology factor) by 168 percent. In the case of phosphate, an important water pollutant, emissions into surface waters increased by 1,845 percent between 1946 and 1968, while population rose by 42 percent, the amount of cleansers per capita remained constant, and the amount of phosphate per unit amount of cleansers increased by 1,270 percent due to the technology factor: the introduction of phosphate-containing detergents in place of soap. Similarly, between 1946 and 1967, when nitrogen oxides emitted by cars increased by 628 percent, population rose by 41 percent, vehicle miles per capita doubled, and nitrogen oxides emitted per mile increased by 158 percent.

It is apparent, then, that in the United States the factor most responsible for the sharp increases in pollutants since World War II—and the factor most capable of reducing pollution—is production technology: the new methods used to produce vehicular travel, cleansers, crops, beer, and many other goods. It can be argued, of course, that in developing countries the situation is different and that their impact on the environment is in fact largely due to what many people regard as their most prominent feature—rapid population growth. Unfortunately, the available data on pollutant levels in developing countries are scanty and incomplete, so that a numerical analysis such as that described for the United States is impossible. However, the problem can be approached indirectly, based on what is already known about the relation between certain pollutants and the production processes that generate them.

For example, it has been established that the rising levels of nitrate—a pollutant that contributes to eutrophication and to health problems in drinking-water supplies—in U.S. and European surface waters is largely due to the application of nitrogen fertilizer to crops. Where such data have been obtained, about 20 to 25 percent of the applied nitrogen reaches surface waters. Hence, subject to this range of uncertainty, the amount of nitrogen fertilizer applied to crops can be used, as a proxy, to represent the resultant level of nitrate in surface waters. Thus, the relative effects of the population, affluence, and technology factors on the pollutant, nitrate, can be estimated if, for a given country or area, changes over time in the following factors can be computed:

population; crop production per capita; and nitrogen fertilizer used per unit crop.

Data on these factors for the period 1970–1980 are available for most developing countries; they are conveniently expressed as the annual rate of change. In ninety developing countries, nitrogen fertilizer use (a proxy for nitrate pollution) increased by an average of 8.6 percent per year, while the rise in population averaged 2.5 percent per year, crop production per capita (affluence) decreased by 0.06 percent annually, and fertilizer use per unit of crop production (the technology factor) increased by 6.6 percent. The impact of the technology factor on the amount of nitrogen fertilizer used, and hence on the level of nitrate pollution, considerably outweighs the effect of both the rapidly rising population and "affluence." Similar analyses show that the introduction of automotive vehicles and power plants in developing countries has had a significantly greater impact on the resultant pollution levels than either population or "affluence."

In sum, the data both from an industrial country like the United States and from developing countries show that the largest influence on pollution levels is the pollution-generating tendency of the system of industrial and agricultural production, and the transportation and power systems. In all countries, the environmental impact of the technology factor is significantly greater than the influence of population size or of affluence.

What does this mean for developing countries, where increased production is the engine of economic progress? At present, developing countries usually introduce those technologies that have proven to be both highly productive and ecologically unsound in industrial countries—nitrogen fertilizer, for example. As pointed out in the recent Bruntland report, the upshot is that "the industries most heavily reliant on environmental resources and most heavily polluting are growing most rapidly in the developing world, where there is more urgency for growth and less capacity to minimize damaging side effects."

Thus, especially in developing countries, the question of environmental quality is an inseparable component of the issue of economic development. To claim that the two are in conflict and that environmental quality can only be achieved at the expense of development ignores the dominant role of production technologies in determining environmental impact. Economic development can proceed without a concomitant decrease in environmental quality if it is based on an appropriate, ecologically benign production technology. For example, crop production can be increased without incurring the environmental hazards of conventional chemical agriculture by practicing organic farming instead. The apparent conflict between environmental quality and economic development that motivates proposals to limit the growth of population and/or production can be largely eliminated by the proper choice of production technologies.

What, next, is the evidence that overpopulation is responsible for famine? Hunger is widespread in the world and those who believe that the world's resources are already insufficient to support the world population cite this fact as evidence that the world is overpopulated. Once more, it is revealing to examine the actual data regarding the incidence of malnutrition. It is useful to remember that people in other countries did not go hungry because they sent food to Ethiopia to relieve the famine there—that what was sent to Ethiopia was *surplus* food. In fact, the world produces more than enough food to feed the total world population. Total world production of food, equally distributed to the global population, would today provide everyone with more than enough for the physiologically required diet. According to a recent estimate by the United Nations Food and Agriculture Organization, the world produces enough grain alone to provide every person on earth with 3,600 calories a day—more than one and a half times the calories required in a normal diet. Enough grain is produced to give everyone on earth two daily loaves of bread.

Famine is caused, not by a global food shortage, but by the grossly uneven distribution of the global food supply. This is not an ecological phenomenon but a political and economic one. Neither England nor Haiti produces enough food for its own population, but hunger is much more prevalent in Haiti than it is in England, because Haiti cannot afford to import enough food to make up the deficit, while England can.

Hunger and malnutrition are also a consequence of maldistribution of food *within* a country. From a detailed study of nutritional levels among various populations in India in 1967, we learn, for example, that in Madras State more than one-half the population consumed significantly less than the physiologically required calories and protein in their diet. However, the *average* values for all residents of the state represented 99 percent of the calorie requirement and 98 percent of the protein requirement. What this means, of course, is that a significant part of the population received more than the required dietary intake. About one-third of the population received 106 percent of the required calories and 104 percent of the required protein; about 8 percent of the population received 122 percent of the calorie requirement and 117 percent or more of the protein requirement. These dietary differences were determined by income. The more than one-half of the population that received significantly less than the physiologically required diet earned less than $21 per capita per year, as compared with the statewide average of $33. What these data indicate is that hunger in Madras State, defined simply in terms of a significantly inadequate intake of calories and protein, was not the result of too many people and not enough food. Rather, it resulted from the social factors that govern the distribution of available food—and income—among the population.

Thus, the available data about both hunger and environmental quality in developing countries show that they have been governed less by population size than by the countries' economic status and the kinds of production technology employed. It remains true, nevertheless, as shown by the multiplicative relation among the three factors that govern pollution, that, other things being equal, a rising population will contribute to the demand for food and to environmental stress. Even though the impact of population on environmental quality is less than the effect of the technology of production, in developing countries it is not negligible, and must be taken into account. It is of interest, therefore, to consider what is known about the stabilization of human populations and how that demographic process is related to biological factors such as birth and death rates, and social factors such as economic development.

Like all living things, people have an inherent tendency to multiply geometrically—that is, the more people there are the more people they tend to produce. In contrast, the supply of food rises more slowly, for unlike people it does not increase in proportion to the existing rate of food production. This is, of course, the familiar relationship described by Malthus that led him to conclude that the population will eventually outgrow the food supply (and other needed resources), leading to famine and mass death. The problem is whether other countervailing forces will intervene to limit population growth and to increase food production.

When we turn from merely stating the problem to analyzing and attempting to solve it, the issue becomes much more complex. The simple statement that there is a limit to the growth of the human population, imposed on it by the limited availability of the necessary resources, is a useful but abstract idea. In order to reduce it to the level of reality in which the problem must be solved, we need to analyze the actual relationship between population growth and resources. Current views on this question are neither simple nor unanimous.

One view is that the cause of the population problem is uncontrolled fertility, the countervailing force—the death rate—having been weakened by medical advances. According to this view, given the freedom to do so, people will inevitably produce children faster than the goods needed to support them. It follows, then, that the birth rate must be deliberately reduced to the point of "zero population growth."

The methods that have been proposed to achieve this kind of direct reduction in birth rate vary considerably. One method is family planning: providing people with effective contraception and access to abortion facilities and educating them about the value of having fewer children. Another suggestion, sometimes called the "lifeboat ethic," is to withhold food from the people of starving developing countries which, having failed to limit their birth rate sufficiently, are deemed to be too far gone or too unworthy to be saved. The author of this so-called ethic, Garrett Hardin, stated it this way:

> So long as we nations multiply at different rates, survival requires that we adopt the ethic of the lifeboat. A lifeboat can hold only so many people. There are more than two billion wretched people in the world—ten times as many as in the United States. It is literally beyond our ability to save them all. . . . Both international granaries and lax immigration policies must be rejected if we are to save something for our grandchildren.

But there is another view of population that is much more complex. It is based on the evidence, amassed by demographers, that the birth rate is not only affected by biological factors, such as fertility and contraception, but also by equally powerful social and economic influences. Demographers have delineated a complex network of interactions among the various biological and social factors. It shows that population growth is not the consequence of a simple arithmetic relationship between birth rate and death rate. Instead, there are circular relationships in which, as in an ecological cycle, every step is connected to several others.

Thus, while a reduced death rate does, of course, increase the rate of population growth, it can also have the opposite effect, since families usually respond to a reduced rate of infant mortality by opting for fewer children. This negative feedback modulates the effect of a decreased death rate on population size. Similarly, although a rising population increases the demand on resources, it also stimulates economic activity, which in turn improves educational levels. This tends to raise the average age at marriage and to facilitate contraceptive practices, leading to a reduced birth rate, which mitigates the pressure on resources.

In these processes, there is a powerful social force that reduces the death rate (thereby stimulating population growth) and leads people voluntarily to restrict the production of children (thereby reducing population growth). That force, simply stated, is the quality of life: a high standard of living; a sense of well-being; security in the future. When and how the two opposite effects of this force are felt differs with the stages in a country's economic development. In a premodern society, such as England before the industrial revolution or India before the advent of the English, both death rates and birth rates were high. But they were in balance and population size was stable. Then, as agricultural and industrial production began to increase and living conditions improved, the death rate began to fall. With the birth rate remaining high, the population grew rapidly. However, some thirty to forty years later, as living standards continued to improve, the decline in the death rate persisted, but the birth rate began to decline as well, reducing the rate of population growth.

Swedish demographic data, which are particularly detailed, provide a good example of this process. In around 1800, Sweden had a high birth rate, about 33 per 1,000 population, but since the death rate was equally high, the population was in balance. Then as agriculture and, later, industrial production advanced, the death rate dropped until, by the mid-nineteenth century, it stood at about 20 per 1,000. Since the birth rate remained virtually constant during that period, there was a large excess of births over deaths and the population increased rapidly—an early version of the "population explosion." Then the birth rate began to drop, until in the mid-twentieth century it reached about 14 per 1,000, when the death rate was about 10 per 1,000. Thus, under the influence of a constantly rising standard of living, the population moved, with time, from a position of balance at high birth and death rates to a new position of near balance at low birth and death rates. But in between, the population increased considerably.

This process, the demographic transition, has been characteristic of all industrialized countries. In these countries, the death rate began to decline in the mid-eighteenth century, reaching an average of 30 per 1,000 in 1850, 24 per 1,000 in 1900, 16 per 1,000 in 1950, and 9 per 1,000 in 1985. In contrast, the birth rate remained constant at about 40 per 1,000 until 1850, then dropping rapidly, reaching 32 per 1,000 in 1900, 23 per 1,000 in 1950, and 14 per 1,000 in 1985. As a result, populations grew considerably, especially in the nineteenth century, then slowed to the present net rate of growth of 0.4 percent per year.

The same process has been under way in developing countries, but with a longer time lag between the declines in death rate and birth rate. In developing countries, the average death rate was more or less constant, at about 38 per

1,000 until 1850, then declining to 33 per 1,000 in 1900, 23 per 1,000 in 1950, and 10 per 1,000 in 1985. The average birth rate, on the other hand, remained at a constant high level, 43 per 1,000, until about 1925; it has since declined at an increasing rate, reaching 37 per 1,000 in 1950, and 30 per 1,000 in 1985. As a result, the increase in the population of the developing countries that began around 1850 has started to slow down and those countries' populations are now growing at an average rate of about 1.74 percent annually. It is important to note that the *death rates* of developed and developing countries are now nearly the same and, given the inherent biological limits, are not likely to decline much further. Thus, in developing countries the progressively rapid drop in birth rate will accelerate progress toward populations that, like those of developed countries, are approximately in balance.

One indicator of the quality of life—infant mortality—is especially decisive in this process. Couples respond to a low rate of infant mortality by realizing that they no longer need to have more children to replace the ones that die. Birth control is, of course, an essential part of this process; but it can succeed—barring compulsion—only in the presence of a rising standard of living, which generates the necessary motivation. There is a critical point in the rate of infant mortality below which the birth rate begins to drop sharply, creating the conditions for a balanced population. This process appears to be just as characteristic of developing countries as of developed ones. Thus, where infant mortality is particularly high, as in African countries, the birth rate is also very high. Infant mortality is always very responsive to improved living conditions, especially with respect to nutrition. Consequently, there is a kind of critical standard of living which, if achieved, can lead to a rapid reduction in birth rate and an approach to a balanced population.

Thus, in human societies, there is a built-in process that regulates population size: if the standard of living, which initiates the rise in population, continues to increase, the population eventually begins to level off. The chief reason that populations in developing countries have not yet leveled off is that this basic condition has not yet been met. The explanation is a fact about developing countries that is often forgotten: they once were, and in the economic sense often still remain, colonies of more developed countries. In the colonial period, western nations introduced improved living conditions (roads, communications, engineering, agricultural and medical services) as part of their campaign to increase the labor force needed to exploit the colony's natural resources. (The anthropologist Clifford Geertz has pointed out, for example, that in Indonesia Dutch colonists imposed a tax on the Indonesian population that could only be paid in labor.) This increase in living standards initiates the first phase of the demographic transition; death rates fall, but with birth rates remaining high, there is a rapid increase in population. However, since most of the resultant wealth does not remain in the colony, the second (or population-balancing) phase of the demographic transition is hindered. Instead, the wealth produced in the colony is largely diverted to the advanced country—where it helps that country achieve for itself the second phase of the demographic transition. Thus, colonialism is a kind of demographic parasitism: the second, population-balancing phase of the demographic transition in the colonialist country is fed by the suppression of that same phase in the colony.

The colonies, whether governed legally (albeit after military conquest) or—as in the case of the U.S. control of Latin American countries—by extralegal and economic means, have now become the developing countries of the Third World. Their characteristic condition—large and rapidly growing populations; grinding poverty; desperate efforts for economic development, now hampered by huge debts—is not the outcome of a "primitive" past. The Eskimo peoples are an illuminating test case, for like African countries, let us say, they are "undeveloped" by modern industrial standards. Yet having never been colonized, the Eskimo lands, unlike Africa, show no signs of a "population explosion."

In sum, as the demographer Nathan Keyfitz has concluded, in the period 1800–1950 colonialism resulted in the development of an excess of 1 billion in the world population, largely in the tropics.

Given this background, what can be said about the various alternative methods of achieving a stable world population? In India there has been an interesting, if partially inadvertent, comparative test of two possible approaches: family-planning programs, and efforts, also on a family basis, to elevate the living standard. The results of this test show that while the family-planning effort itself failed to reduce the birth rate, improved living standards succeeded.

In 1954 a Harvard team undertook the first major field study of birth control in India. The population of a number of test villages was provided with contraceptives and suitable educational programs. Over a six-year period, 1954–60, birth rates, death rates, and health status in this population were compared with the rates found in an equivalent population in villages not provided with the birth control program. A follow-up in 1969 showed that the population control program had failed. Although in the test population the birth rate dropped from 40 per 1,000 in 1957 to 35 per 1,000 in 1968, a similar reduction also occurred in the comparison population. The birth control program had no measurable effect on the birth rate.

We now know why the study failed, thanks to a remarkable book by Mahmood Mamdani, *The Myth of Population Control.* He investigated in detail the impact of the study on one of the test villages, Manupur. What Mamdani discovered confirms the view that population control in a country like India depends on the *economic* factors that indirectly limit fertility. Talking with the Manupur villagers, he discovered why, despite the study's statistics regarding ready "acceptance" of the offered contraceptives, the birth rate was not affected:

> One such "acceptance" case was Asa Singh, a sometime land laborer who is now a watchman at the village high school. I questioned him as to whether he used the [birth control] tablets or not: "Certainly I did. You can read it in their books. From 1957 to 1960, I never failed." Asa Singh, however, had a son who had been born sometime in "late 1958 or 1959." At our third meeting I pointed this out to him. . . . Finally he looked at me and responded. "Babuji, someday you'll understand. It is sometimes better to lie. It stops you from hurting people, does no harm, and might even help them." The next day Asa Singh took me to a friend's house . . . and I saw small rectangular boxes and bottles, one piled on top of the other, all arranged as a tiny sculpture in a corner of the room. This man had made a sculpture of birth control devices. Asa Singh said: "Most of us threw the tablets away. But my brother here, he makes use of everything."

Such stories have been reported before and are often taken to indicate how much "ignorance" has to be overcome before birth control can be effective in countries like India. But Mamdani takes us much deeper into the problem, by asking why the villagers preferred not to use the contraceptives. In one interview after another, he discovered a simple, decisive reason: in order to advance their economic condition, to take advantage of the opportunities newly created by the development of independent India, *children were essential.* Mamdani makes this very explicit:

> To begin with, most families have either little or no savings, and they can earn too little to be able to finance the education of *any* children, even through high school. Another source of income must be found, and the only solution is, as

> one tailor told me, "to have enough children so that there are at least three or four sons in the family." Then each son can finish high school by spending part of the afternoon working. . . . After high school, one son is sent on to college while the others work to save and pay the necessary fees. . . . Once his education is completed, he will use his increased earnings to put his brother through college. He will not marry until the second brother has finished his college education and can carry the burden of educating the third brother.

Mamdani points out that "it was the rise in the age of marriage—from 17.5 years in 1956 to 20 in 1969—and not the birth control program that was responsible for the decrease in the birth rate in the village from 40 per 1,000 in 1967 to 35 per 1,000 in 1968. While the birth control program was a failure, the net result of the technological and social change in Manupur was to bring down the birth rate."

Here, then, in the simple realities of an Indian village are the principles of the demographic transition at work. There is a way to control the rapid growth of populations in developing countries. It is to help them develop, and to achieve more rapidly the level of welfare that everywhere in the world is the real motivation for reducing the birth rate.

Against this conclusion it will be argued, to quote Hardin, that "it is literally beyond our ability to save them all." This reflects the view that there is simply insufficient food and other resources in the world to support the present world population at the standard of living required to motivate the demographic transition. It is sometimes pointed out, for example, that the United States consumes about one-third of the world's resources to support only 6 percent of the world's population, the inference being that there are simply not enough resources in the world to permit the rest of the world to achieve the standard of living and low birth rate characteristic of the United States.

The fault in this reasoning is readily apparent from the

actual relationship between the birth rates and living standards in different countries. The only available comparative measure of standard of living is the gross national product (GNP) per capita. Neglecting for this purpose the faults inherent in GNP as a measure of the quality of life, a plot of birth rate against GNP per capita is very revealing. For example, in 1984 in the United States GNP per capita was $15,541 and the birth rate was 16 per 1,000. In the poorest countries (GNP per capita less than $500 per year), the birth rates were 32 to 55 per 1,000. In those countries where GNP per capita was $4,000 to $5,000 (for example, Greece), the birth rate ranged from 15 to 19 per 1,000. Thus, in order to bring the birth rates of the poor countries down to the low levels characteristic of the richer ones, the poor countries do not need to become as affluent (at least as measured by GNP per capita) as the United States. By achieving a per capita GNP only, let us say, one-third of that of the United States, these countries could reach birth rates almost as low as those of the European and North American countries.

In a sense, the demographic transition is a means of translating the availability of a decent level of resources, especially food, into a voluntary reduction in birth rate. The per capita cost, in GNP, of increasing the standard of living of developing countries to the point that would motivate a voluntary reduction in birth rate is small, compared to the wealth of the rich, developed countries—a much neglected, global bargain.

I have tried, thus far, to analyze the Third World problem based on data about the demographic changes that have already taken place. The data confirm what the world already painfully knows: that in comparison to developed countries, the Third World is terribly impoverished; that it is struggling, against great odds, to develop and to increase its peoples' standard of living; that because the rate of economic development has barely kept up with the rapidly rising population, the standard of living (as measured by

GNP per capita) has remained essentially the same in recent years.

Proposals on how to resolve this network of intransigent problems are heavily conditioned by fear of the future. What will the world be like fifty or a hundred years from now if Third World populations continue to grow rapidly, vastly increasing the ranks of the poor and intensifying the environmental impact of our ecologically unsound technologies? Present demographic trends indicate that developing countries as a whole will reach the death rate now characteristic of developed countries in about 2000; about thirty or forty years later, they will reach the birth rate now characteristic of developed countries. This suggests that perhaps four to five decades from now, developing countries, on the average, will have achieved the sort of demographic stability now found in developed countries. An estimate by the World Bank, based on earlier trends, is less optimistic: approximate stability will be reached about a hundred years from now, when the world population will be about 10 billion, or about twice its present size.

World food production is now well above the minimum requirement of the present world population, and growing about 30 percent faster than the population. If present trends continue, there will be more than enough food to support a world population of 10 billion when that relatively stable size is reached. At least in these rudimentary terms, the developing countries could traverse the demographic transition without encountering the massive famines predicted by the adherents of the "carrying capacity" hypothesis. But if nothing is done to realign the distribution of food and economic resources, the world will then have twice as many poor and hungry people as it does now. And in that time total production in developing countries will have quadrupled or more so that they, rather than the countries of the north, will then account for most of the world's pollution. Thus, if the world continues on its present path, the moral concerns engendered by massive poverty, and the practical concerns about the degradation of the environment, will only intensify. In sum, the Third World problem will still be with us, only larger and more devastating than before.

Clearly, something more must be done. The pressure on food supplies, resources, and the environment would be reduced if birth rates were to decline faster than they are at present. Improvement in the standard of living—and hence a faster decline in the birth rate—would be hastened if the rate of economic development could be accelerated. Increased impact on resources and the environment could be avoided if development were based on ecologically sound technologies of production.

All of these problems have a common solution: the elimination of poverty. Poverty is the reason for the failure thus far of developing countries to stabilize their populations. Poverty is the reason why their peoples are malnourished, sick, and hungry. Poverty is the reason why they experience such difficulty in applying the remedy: ecologically sound economic development. Poverty engenders poverty, holding the efforts of developing countries to overcome its tragic effects in a tight, nearly incapacitating embrace.

This is the distant outcome of colonial exploitation. Colonialism has determined the distribution of both the world's wealth and its human population, accumulating most of the wealth north of the equator and most of the people below it. The only remedy, I am convinced, is to return some of the world's wealth to the countries whose resources and peoples have borne so much of the burden of producing it—the developing nations. Such colonial reparations ought to be paid not only in goods but, more usefully, in the means of producing them. And the productive processes should be those that correct both the environmental and economic defects of the technologies that have enveloped the global ecosphere in pollution. Obviously, this proposal would involve exceedingly difficult economic, social, and political problems, especially for the rich countries. But the alternative solutions thus far advanced are at least as difficult and socially stressful; and some are morally repugnant.

A major source of confusion is that the diverse proposed solutions to the population problem, which differ so sharply in their moral postulates and their political effects, appear to have a common base in scientific fact. It is, after all, equally true that the size of the population can be reduced by promulgating contraceptive practices (providing they are used), by elevating living standards, or by withholding food from starving nations. But what I find particularly disturbing is that behind the screen of confusion between scientific fact and political intent there has developed an escalating series of what can be only regarded as inhumane, abhorrent political schemes put forward in the guise of science. There have been "triage" proposals that would condemn whole nations to death through some species of global "benign neglect." There have been schemes for coercing people to curtail their fertility, by physical and legal means that are ominously left unspecified. Now we are told that we must curtail rather than extend our efforts to feed the hungry peoples of the world. Where will it end? Is it conceivable that the proponents of coercive population control will be guided by one of Garrett Hardin's earlier, astonishing proposals:

> How can we help a foreign country to escape overpopulation? Clearly the worst thing we can do is send food. . . . Atomic bombs would be kinder. For a few moments the misery would be acute, but it would soon come to an end for most of the people, leaving a few survivors to suffer thereafter.

The present confusion can be remedied by recognizing all of the proposals for what they are—not scientific observations but value judgments that reflect sharply differing ethical views and political intentions. The family-planning approach, if applied as the exclusive solution to the population problem, would put the burden of remedying a fault created by a social and political evil—colonialism—solely on the individual victims of that evil. The so-called "lifeboat ethic" would compound the original malevolence of colonialism by forcing its victims to forgo the humane course toward a balanced population—improvement of living standards—or, if they refuse, to abandon them to destruction, or even to thrust them toward it.

My own purely personal conclusion is, like all of these, not scientific but political: the world population crisis, which is the ultimate outcome of the exploitation of poor nations by rich ones, ought to be remedied by returning to the poor countries enough of the wealth taken from them to give their peoples both the reason and the resources voluntarily to limit their own fertility.

In sum, I believe that if the root cause of the world population crisis is poverty, then to end it we must abolish poverty. And if the cause of poverty is the grossly unequal distribution of the world's wealth, then to end poverty, and with it the population crisis, we must redistribute that wealth, among nations and within them.

8

ENVIRONMENTAL ACTION

THE CAMPAIGN TO CLEAN up the environment has largely failed; but not for lack of effort. The EPA's massive, largely futile regulatory program to control pollution and the responses required of the enterprises that produce it are only part of the work done in the name of restoring environmental quality. A remarkable array of public organizations are involved as well. These range in size and financial resources from the National Wildlife Federation—one of some dozen environmental organizations headquartered in Washington, D.C., which in 1988 had over a million members and an annual budget of $67 million, to Concerned Citizens of Cohasset (Massachusetts), to name one of perhaps five thousand similar groups, which consists of a hundred citizens and has a budget of zero. What all these groups have in common is that they have opinions about what is wrong with the environment and try to influence what is done to improve it. But beyond that common aim, they differ considerably: in their perception of the environmental crisis; in what they think ought to be done about it; in the methods they use to carry out their aims; and in the effectiveness of their efforts. They differ especially in the way they relate to the largely failed effort, since 1970, to carry out the purpose of the National Environmental Policy Act: "to prevent and eliminate damage to the environment and the biosphere."

That the achievement of this aim—prevention—depends on social intervention in the hitherto private governance of production technology helps to explain why the considerable effort to improve the environment has thus far had so little effect. This effort, however forceful, has thus far met a politically immovable object: the conviction, deeply embedded in American society, that the decisions that determine what is produced and by what technological means ought to remain in private, generally corporate hands. In the United States, there appears to be a powerful taboo against even public discussion, let alone criticism, of this basic precept. Environmental organizations must either confront this barrier or find some easier path of action.

One result is that many environmental organizations have been drawn toward explanations of the environmental crisis that avoid its origin in so controversial an arena as economics and that appeal instead to the less contentious principles of ecology. Despite the evidence that the environmental crisis originates in the technosphere, they seek its solution in the ecosphere. Thus, as we have seen, on the assumption that the global ecosphere is a closed system of limited capacity that is stressed by the growing world population, some environmentalists believe that population control is the key to environmental quality. Similarly, that ecological cycles appear to be closed and self-sustained suggests to others that people should also abide by this rule, and live in communities that are supported by their own local or regional resources. Finally, some derive from these same considerations the conclusion that ecological improvement calls upon the human species to abdicate dominion over the rest of nature; to recognize in particular the rights of other animal species (usually limited to furry vertebrates); and, often enough, to reinforce this renunciation with a spiritual commitment to the primacy of the planet—sometimes exemplified by a goddess, Gaia—over people.

It is useful to examine how these different approaches contribute to the environmental effort. Are they really well grounded in the principles of ecology? Do they promise to improve the mediocre environmental record? How do they relate to the government's failing efforts?

Many people have been attracted to the environmental movement because of their dismay over the state of human society. Some of them are interested not only in correcting the environmental abuses that erode the quality of life but also in establishing a new, happier way of living. They want to create better places to live and better ways to work. They regard industrial society as the root of not only environmental problems but of the social, political, and economic ones as well. They look toward ecology as a guide to the creation of a new culture, a new style of living which, according to Kirkpatrick Sale, a prominent exponent of this view, "is rooted in the natural world, in harmony with natural systems and rhythms, constrained by natural limits and capacities and developed according to the natural configurations of the earth and its inherent life forms."

Seeking to establish an ecologically harmonious way of life—at least for themselves—some people return to the land, growing their own food, building their own house and furniture, recycling their waste. Others have a wider vision, seeking "self-reliance, not so much at the individual as at the regional level." This approach, "bioregionalism," envisions a society based on ecologically defined regions, rather than on areas specified by their political boundaries. It includes the belief that such an ecologically founded society will have many laudable features: it will be cooperative rather than competitive, decentralized rather than centralized, interdependent rather than polarized, "evolving" rather than growing, peaceful rather than violent.

Certainly, it makes ecological, and, as we have seen, ultimately economic sense to organize society in harmony with nature. Problems arise, however, when one attempts to translate this general proposition into practice. According to Sale, bioregionalism is achievable because it has an appeal that overrides conventional political distinctions, thereby avoiding polarization and opposition. "The bioregional idea has the potential to join what are traditionally thought of as the right and the left in America because it is built on and appeals to values that, at bottom, are shared on both sides."

Based on this benign outlook, Sale expects bioregionalism to remake society without encountering opposition from the powerful economic forces that now largely govern it. This would be a remarkable political conjuring trick, for there is no way to reorganize society along ecologically sound lines without challenging head-on the powerful, politically conservative forces—more plainly speaking, the corporations—that now control the system of production. This, as we have seen, is the basic conclusion that emerges from the effort to deal with environmental pollution over the last two decades.

The illusion that environmental improvement is a politically neutral issue is not new. When environmental quality first emerged as a public concern in 1970, a major politician declared that "ecology has become the political substitute for the word 'motherhood.' " The theory that conservatives and liberals will happily join a common campaign to develop an ecologically sound society has been tested by an eminent practitioner of political reality, ex–President Nixon. He learned from experience that the theory is wrong, when he discovered that even his administration's modest environmental efforts were a serious threat to corporate power.

In his 1970 State of the Union message, President Nixon adopted the environmental issue as his own, promising a vigorous campaign to clean up pollution. He said that "the 1970's absolutely must be the years when America pays its debt to the past by reclaiming the purity of its air, its waters and our living environment. It is literally now or never." In his 1971 message, Mr. Nixon, still sounding like a crusading ecologist, declared that environmental quality would be the "third great goal" of what he called the "New American Revolution." But less than a year later, Mr. Nixon had learned that environmentalism is a threat to the basic precepts of corporate power. In September 1971, Nixon announced that he had made a complete reappraisal of his environmental policies and told an audience of auto industry executives that he would not permit environmental concern "to be used sometimes falsely and sometimes in a demagogic way to destroy the system." Nixon had a better appreciation than Sale of the political impact of ecology.

Solving the environmental crisis—as distinct from somewhat diminishing its effect—is fundamentally a political problem because it calls for the establishment of a new, social form of governance over decisions that are now exclusively in private, corporate hands. But until recently, most environmental groups merely debated how the government ought to regulate whatever production facilities the corporate decision makers in their wisdom decide to build and operate. Since pollution is inherent in the very design of the production technologies, once the technological choice is made, regulation can have only a limited effect, addressing the symptom instead of the disease.

In alluding to the environmental impact of production technologies, environmentalists sometimes speak of the "hard path" and the "soft path." The hard path—the huge, centralized technologies created without concern about their environmental impact—leads to nuclear power, chemical agriculture, and high-powered cars. The soft path leads to technologies more appropriate in scale and design to their human purpose and ecological setting: cogenerators, solar energy, organic farming, small cars, mass transit, and bicycles. This distinction is useful, but it is incomplete; in fact, the road to the soft path of ecologically appropriate technology is a hard, political one.

There is in fact a "hard path" and a "soft path" in environmental politics. The soft path is the easy one; it accepts the private corporate governance of production decisions and seeks only to regulate the resultant environmental impact. (And free to do so, the corporations have invariably chosen the "hard" technologies, which are so impervious to environmental control.) In environmental politics, the hard path is the difficult one; it would confront the real source of environmental degradation—the technological choice—and debate who should govern it, and for what purpose. The hard political path is the only workable route to the soft environmental path.

In the United States, the major Washington-based environmental organizations are aware that in order to influence the resolution of environmental issues, they must find a way to participate in the political process. But they have generally chosen to travel the soft political path. This was evident in *An Environmental Agenda for the Future,* published in 1986 by a group of ten self-described "leaders of America's foremost environmental organizations." The Agenda recognizes the need to confront the *cause* of environmental degradation, stating that "because the laws and regulations often have not dealt with root causes, they have been inadequate to cope with added problems that have arisen, partly from new technologies."

Given this laudable intention, and what we now know about the origin of environmental pollution, one would expect the environmental leaders to guide us toward the future by proposing national policies such as the following: the present reliance on nonrenewable energy sources, especially nuclear power, should be replaced by renewable solar energy systems; the present chemically based system of agriculture should be replaced by organic farming; the inherently dangerous petrochemical industry should be rolled back, its products and processes replaced with ones more compatible with the environment; the nation's present heavy dependence on cars and trucks should be replaced by a vast expansion of railroads and mass transit systems.

However, the Agenda offers a policy statement only with respect to its support for population control: "The Administration should establish formal population policies, including goals for the stabilization of population at a level that will permit sustainable management of resources and a reasonably high quality of life for all people."

Apparently, the environmental leaders believe that overpopulation, but not the technology of production, constitutes a "root cause" of environmental degradation and thus warrants a national policy. Indeed, the Agenda approaches very gingerly any measure that calls for socially mandated changes in the technology of production. For example, according to the Agenda, organic farming should be "researched and encouraged," but no political action is proposed to actually put this advantageous technology into practice.

The Agenda program appears to reflect the environmental leaders' effort to adjust the goals of environmentalism to the reality of Washington politics. As they see it, the reality is this: Legislation to improve environmental quality, for example by enacting more stringent pollution standards, means that a majority of Congress must be won over. But legislators are also under the influence of well-financed industrial lobbyists who fiercely resist such restrictions on their companies' operations. In the past, these battles have often been lost by the environmentalists or, at best, have produced a legislative compromise. The alternative to lobbying is litigation, a lengthy, costly, and often inconclusive process. Thus far, as we have seen, such efforts have not done much to improve the state of the environment.

With the Reagan administration in power, which strived for minimalist government programs in all but military affairs, environmental lobbying seemed particularly fruitless. The environmental organizations found themselves facing mounting costs and diminishing returns. This may explain a tendency on their part to seek less frustrating routes toward environmental improvement. The main outcome is the formation of small committees, composed of leaders of the Washington-based environmental organizations and corporate executives, that seek to work out compromise positions, thus avoiding the cost, delay, and frustration involved in lobbying and litigation.

A good example is the Acid Rain Roundtable, a twelve-member group representing environmental organizations and power companies. The Roundtable reported the results of a study they commissioned to devise "cost-effective programs" that reduce emissions of sulfur dioxide and nitrogen oxides. The study concluded that a reduction of 26 percent in sulfur dioxide emissions and 20 percent in nitrogen oxide emissions could be achieved at an annual cost of about $2.6 billion. The report opposed the strategy of reducing sulfur dioxide emissions by installing scrubbers, which were regarded as too costly. According to the study, the most cost-effective solution is to cut back the burning of high-sulfur coal—a step that would reduce coal mining in the Midwest and northern Appalachia. With respect to the resultant unemployment, one Roundtable participant noted that the study was directed toward ways "to address miner unemployment by means other than mandating the installation of scrubbers." The report proposed to aid the miners who would lose their lifelong jobs by helping them obtain unemployment and health insurance for one year after they are thrown out of work. Regardless of its merits, the basic structure of the compromise proposed by the study is clear: it relieves the power companies of the cost of installing scrubbers, shifting the economic consequences of achieving a modest reduction in emissions onto the miners.

According to one Roundtable member, the study was designed to "break the legislative deadlock." This may explain why mine workers were excluded from the Roundtable negotiations, for as important parties to the issue, they would receive respectful attention if it were resolved by conventional democratic means, in the legislature.

It can be argued, of course, that given their considerable power, it is better to compromise with the corporate polluters in the hope of making some partial improvement than to engage them in a battle that may only end in frustration. On the other hand, there is a danger that in the course of negotiating a compromise, the environmental organizations will become hostage to the corporations' power and experience the "Stockholm syndrome," in which hostages take on the ideology of their captors. This may explain why Jay D. Hair, executive director of the National Wildlife Federation, which is heavily involved in "responsible dialogue with corporate leaders," has admonished environmentalists that "our arguments must translate into profits, earnings, productivity and incentives for industry"—precisely the arguments of corporate executives against the regulation of pollution.

Recently, this affinity for corporate interests has taken a practical turn in the form of corporate sponsorship of conservation projects undertaken by another Washington-based organization, the World Wildlife Fund (WWF). A WWF brochure soliciting such sponsorship points out: "Your company can use a World Wildlife Fund tie-in to achieve virtually every objective in your marketing plan:

- New product launches
- Corporate awareness
- New business contacts
- Brand awareness
- Brand loyalty. . . ."

As a prominent example of this approach, the brochure displays a photograph of a Jaguar car next to one of the animal of the same name and a caption that reads, in part: "To help insure the survival of this powerful feline, the inspiration for its elite line of luxury cars, Jaguar Cars Inc. has committed to a three-year WWF program to support the Cockscomb Jaguar Preserve in Belize."

The caption does not point out that while the real jaguar operates very efficiently on solar energy, the vehicular Jaguar burns a nonrenewable, highly polluting fuel and does so very inefficiently.

One result of such organizations' tendency to travel the soft path of environmental politics has been the appearance of a new "grass-roots" environmental movement that is prepared to take the hard political path. This movement has arisen out of the direct impact of environmental pollution on community life. It is chiefly concerned with toxic chemicals, toxic waste dumps, and dioxin-producing trash incinerators. It is exemplified by Love Canal, where the local residents organized to demand the actions that finally led to the evacuation of the area and some compensation for their financial losses. Lois Gibbs, who led that fight, has formed a national federation of similar community groups, the Citizen's Clearinghouse for Hazardous Wastes. Another organization, the National Toxics Campaign, also helps citizens' groups understand the hazards from local sources of chemical pollution and how to act to prevent them. Internationally, as well as in the United States, Greenpeace has organized dramatic, frequently daring, demonstrations against nuclear bomb tests, marine environmental hazards, trash-burning incinerators, and sources of toxic chemicals. The Public Interest Research Groups (PIRGs) organized by Ralph Nader are also part of this informal coalition.

For such groups, the front line of the battle against chemical pollution is not in Washington but in their own community. For them, the issues are clear-cut and not readily compromised: A waste management company decides to build a trash-burning incinerator, but the community, fearful of the health effects, doesn't want it. A chemical company keeps dumping toxic wastes in a leaky lagoon, but the community wants the practice stopped and the lagoon cleaned up. In these battles, there is little room for compromise; the corporations are on one side and the people of the community on the other, directly challenging the corporation's exclusive power to make decisions that threaten the community's health.

Thus, confronted by powerful corporate opposition, the environmental movement has split in two. The older national environmental organizations, in their Washington offices, have taken the soft path of negotiation, compromising with the corporations about how much pollution is acceptable, and sometimes helping to market their products, even when they are ecologically unsound. The people living in the polluted communities have taken the hard path of confrontation, demanding not that the dumping of hazardous waste be slowed down, but stopped; not that dioxin-producing incinerators be equipped with unworkable emission controls, but abandoned in favor of recycling. The national organizations deal with the environmental disease by negotiating about the kind of "Band-Aid" to apply to it; the community groups deal with the disease by trying to prevent it.

As a result, the conflicts that have been such a notable feature of the environmental issue have divided environmentalists into two different arenas. In Washington's congressional hearing rooms, as they have from the early 1970s, standards for the pollutants now subject to regulation—and for the more numerous unregulated ones—are disputed. Here the antagonists are usually the EPA, the lobbyists employed by the national environmental organizations, and those hired to serve the corporations. They battle over the seriousness of a pollutant's effects, the desirability of setting a particular standard of exposure, and the feasibility of employing various control devices for that purpose. The battle lines are predictable: the environmental organizations call for stricter standards, the corporate lobbyists for weaker ones, with the EPA reflecting whatever position seems politically expedient at the moment. In recent years, the expedient position has often been identical with the corporations'. However rich in technical justification, noble in social purpose, or adamant in defense of corporate profits, these arguments, which usually end in a compromise, are fruitless. Regardless of their outcome, as we have seen, they have had little impact on environmental quality.

The other environmental arena is far removed from Washington's well-kept conference chambers. In Washington pollutants are discussed; in this other arena they are experienced. The ambiance is characterized by foul air, dingy water, and sick and dying people. These battles are fought in thousands of different places: at toxic dumps, trash-burning incinerators, and chemical plants. The antagonists are the corporate polluters on one side and the people of the local community on the other. The polluters are directly confronted by the polluted, from children with birth defects at Love Canal to householders with dirty laundry downwind from an incinerator. Banded together in an ad hoc committee under an acronym such as STOP, RAGE, or WASTE, the local residents fight not for improved standards of exposure but for no exposure at all. Their preferred standard of exposure is zero. They want pollution prevented rather than controlled.

Some of these battles take place in supermarkets, where

shoppers reject plastic bags in favor of paper ones, and stop buying products that are notoriously polluted. A recent example is the project, organized by the National Toxics Campaign, that has thus far persuaded 1,200 supermarkets to stop selling produce treated with cancer-causing pesticides.

What distinguishes the grass-roots campaigns from the activities of the older, Washington-based organizations is their environmental strategy. The major federations of grass-roots citizens' committees—the Citizen's Clearinghouse for Hazardous Wastes, the National Toxics Campaign, the PIRGs, and Greenpeace—in confronting the polluters are committed to a strategy of prevention. The old-line Washington-based organizations, in arguing—or negotiating—for standards, follow a strategy of control.

For several years this divided scenario has dominated environmentalism in the United States. But then in January 1988, not by coincidence the end of the Reagan administration, the EPA's bureaucratic machinery seemed to take a sharp, unanticipated ideological turn, which is bound to rearrange the battle lines. It is best appreciated against the background of the agency's history. The EPA began its work in the glow of enthusiasm generated by the lively public manifestations of Earth Day 1970. But encumbered by regulatory machinery that conflicts with the original legislative purpose—to "prevent or eliminate damage to the environment"—EPA became enmeshed in the increasingly futile effort to deal with the symptoms rather than the origin of the disease: the promulgation of standards that validate certain, hopefully innocuous, levels of pollution; a series of balancing acts to provide scientific (and often *pseudo* scientific) justification for the standards; protracted litigation over their validity.

During the Reagan administration, lax enforcement and the failure inherent in the EPA's regulatory mandate was compounded by the president's obsession with private enterprise ideology. This apparently led him to appoint administrators—Anne Gorsuch of the EPA and James Watt of the Interior Department—who appeared to be less interested in protecting the environment than the corporations that led the attack on it. Reagan's ideological bias was directly expressed in his order to the OMB to base regulatory standards on "economic efficiency," generating pressure on the EPA to relax their assessments of the risk represented by carcinogens and other toxic environmental agents. This intrusion into the EPA's responsibilities was made worse by the spurious use of "science" to justify it. As a result, one of the legacies of the Reagan era is a considerable decline in public esteem for the EPA and frustration among its personnel.

The great majority of the EPA staff are not White House appointees, but civil servants devoted to the agency's unfulfilled task. Perhaps even more than outside environmentalists, they were frustrated by the environmental failure and often outraged by the OMB's efforts to undermine their work in the name of Mr. Reagan's ideological obsession. That may explain why in January 1988, when in response to an invitation from the EPA's Office of Toxic Substances and the Air and Radiation Program, I arrived to deliver an address on "The Environmental Failure," the auditorium and several TV-linked rooms were packed. And after spelling out the scope and significance of the nearly twenty-year failure to clean up the environment and the need for prevention rather than control, I ended with a call for "an open public discussion of what had gone wrong and why," I was astonished—as Philip Shabecoff, the *New York Times* environmental correspondent, put it—to be "showered by wild applause."

One year later, the evidence that the EPA staff understood that the agency's regulatory program was headed in the wrong direction surfaced in an even more surprising manner. There appeared in the *Federal Register* a formal

"Pollution Prevention Policy Statement" signed by the administrator, Lee M. Thomas, on his last day in office. Apart from the necessary bureaucratic ambiguities, the statement accepted the argument that control measures had failed to satisfactorily improve the environment and that only prevention can succeed. The statement acknowledged that much of EPA's past effort "had been on pollution control rather than pollution prevention.... EPA realizes that there are limits as to how much environmental improvement can be achieved under these [i.e., control] programs, which emphasize management after pollutants have been generated." The statement cited the rapid removal of lead from gasoline as a laudable effort "to reduce pollution at the source."

Thomas's statement also boldly renounced the philosophy of "acceptable exposures," asserting that "today's notice commits EPA to a preventive program to reduce or eliminate the generation of potentially harmful pollutants."

On its face, Mr. Thomas's statement represents a sharp change in EPA policy. If implemented, the statement amounts to a revolution in environmental regulation, at least as practiced until now by both EPA and the Washington-based environmental organizations. Published on the last day of the Reagan administration—and of Mr. Thomas's term of office—the statement was a challenge to George Bush, the new self-anointed "environmental president" and to his EPA administrator, William K. Reilly, who already wore the mantle of "a leading environmentalist" by virtue of his previous presidency of the World Wildlife Fund.

This engaging triad of circumstances engendered a good deal of excitement at EPA headquarters in the first weeks of the Bush administration. Soon word circulated that pollution prevention was at the top of Mr. Reilly's agenda. In May he publicly embraced the Thomas statement, declaring in testimony before a subcommittee of the House Committee on Energy and Commerce that

> there is a growing recognition that traditional approaches—which stress treatment and disposal *after* pollution has been generated—have not adequately dealt with existing environmental problems. Nor will they provide an adequate basis for dealing with emerging problems such as global warming, acid rain, and stratospheric ozone depletion. EPA believes that further improvements in environmental quality will be best achieved by *preventing* the generation of pollutants that may be released to the air, land, and water by eliminating or reducing them at their source and encouraging environmentally safe recycling of those which cannot be eliminated. . . . Controlling pollution is not enough; we will be environmentally and economically better off when pollution is prevented from occurring in the first place. EPA's mandate, in its broadest definition, is to reduce risk—to protect human health, the ecological universe, and natural resources. I believe that pollution prevention along with environmentally safe recycling can and will serve as the primary means of fulfilling this mandate.

By June, EPA gossip raised pollution prevention to the most exalted level of bureaucratic approval: Mr. Reilly's position, it was said, had the full blessing of the president. This was apparently confirmed on June 7, when Mr. Bush spoke to a Washington audience about his environmental program. According to the *Washington Post,* he declared himself not only an environmentalist but a preventionist as well: "His goal will be preventing, not just cleaning up, environmental problems."

Thus, less than six months after Mr. Thomas's statement fired the first, unexpected shot in the apparent regulatory revolution, the insurgent—pollution prevention—had apparently won, holding commanding positions in both the White House and EPA headquarters. To veterans of the environmental movement, the situation was reminiscent of an earlier ideological shock, when Richard Nixon—not known for his interest in government regulation—surprised us all by devoting his first State of the Union message (and the second one as well) to a resounding commitment to environmental quality. Mindful of the sudden collapse of Mr. Nixon's environmental resolve in the face of corporate opposition, I waited for the first confrontation between the administration's new ideology and the political reality: that the policy called for an invasion of the sacred precincts of the corporate board room.

The test was not long in coming. It arose out of the built-in conflict between trash-burning incinerators and recycling. Both the Thomas pollution prevention statement and Mr. Reilly's congressional testimony emphasize that recycling is an important aspect of the new prevention policy. In contrast, a trash-burning incinerator is a control device; it is a means of treating trash, after this pollutant has been generated, in an effort to reduce its environmental impact. Incineration itself involves a series of controls: on the incinerator stack emissions; on the landfill to which the residual ash is consigned; and on the landfill leachate. By converting components that would otherwise become trash into useful materials, recycling prevents all these pollution problems and eliminates the need for such controls. Moreover, because some 80 percent of the trash components can be either burned or recycled, incineration and recycling are inherently incompatible.

The pollution prevention issue arose in a novel way during the course of a dispute over licensing a proposed trash-burning incinerator in Spokane, Washington. Opponents argued that according to the Clean Air Act, the facility must employ the "best available control technology" (BACT). This provision was not written with trash-burning in mind, but rather to facilitate the use of techniques that remove sulfur from coal, thereby reducing sulfur dioxide emissions (and the resultant acid rain) when this fuel is burned. Regardless of its original intent, once expressed in the legal language of the Clean Air Act, BACT is defined as a means of limiting pollutant emissions through "fuel cleaning or treatment." Applied to the trash used to fuel incinerators, this would mean removing and recycling all the components that, when burned, contribute to hazardous emissions. In practice, this would include nearly all of the trash components, and leave very little for the incinerator to burn. In effect, the implementation of BACT would render the incinerator useless. Citing the Thomas statement, EPA Region 10 agreed that BACT was applicable to the Spokane incinerator and referred the issue to the administrator for decision. Mr. Reilly had a momentous opportunity to signal EPA's turn toward prevention by supporting the Region 10 position.

The stakes were very high. If, as Region 10 proposed, instead of allowing the Spokane incinerator to burn all the unseparated trash, it was required, under BACT, to remove pollutants, the "resource recovery" industry would not survive. The only virtue of such an incinerator is that it avoids the difficulties involved in burdening householders with separating their trash and collection crews with the resultant extra work. Deprived of that advantage, incineration becomes even less attractive than it already is by reason of its high cost and environmental hazards. In effect, if Mr. Reilly exercised the BACT provision, he would be not merely prescribing a means of controlling pollution, but dictating the technology of trash disposal. This is, of course, precisely the sort of decision that must be made to prevent pollution: what must be changed is not the system of pollution control, but the very production technology itself, in this case replacing the incinerator with recycling. This fact was laid before Mr. Reilly quite explicitly in a letter from the National Resource Recovery Association (an incinerator-promoting organization that, quite remarkably, is part of a supposedly noncommercial organization, the U.S. Conference of Mayors). Their letter said that the "EPA can lawfully set only emission limits in Spokane's PSD [prevention of significant deterioration] permit; *it cannot prescribe partic-*

ular technologies nor waste management strategies such as recycling to meet those limitations." [emphasis added]

In June 1989, Mr. Reilly rejected the proposal to apply BACT to trash-burning incinerators. He gave no principled reasons for disagreeing with the carefully reasoned Region 10 position. Instead, he fell back on a highly technical point that was, in fact, mistaken. Region 10 had cited studies showing that the removal of certain components from the trash stream reduced the concentrations of toxic metals in incinerator flue gas. However, since the measurements were taken at a point before the gas entered the control device, Mr. Reilly rejected this evidence as not indicative of the actual emissions into the environment. This is a wholly illogical argument, for the control device will only *further* reduce the toxic metal content of the flue gas before it is emitted from the stack.

The Pollution Prevention Policy Statement can be regarded as a kind of bureaucratic last will and testament left by the departing EPA administrator as a guide to his successor. In agreeing to encumber the Spokane incinerator with a recycling requirement that would render it futile, Region 10 was responding to Mr. Thomas's call for "the development of institutional strategies in each of the EPA's media-specific and regional offices to ensure that the pollution prevention philosophy is incorporated into every feasible aspect of internal EPA decision-making and planning." Unfortunately, Mr. Thomas neglected to include EPA headquarters in this admonition.

If the fledgling EPA pollution prevention policy failed to find a nest in Mr. Reilly's office, its treatment at the White House was even worse. There, on June 12, Mr. Bush unveiled his first putative achievement as the new "environmental president": his Clean Air bill. In some respects, the bill does reflect Mr. Bush's praise of pollution prevention a few days earlier, in that it calls for the introduction of new fuels, such as alcohol and methane, that prevent some of the pollution associated with gasoline. But most of the bill, like all of its predecessors, aims at controls and standard setting rather than prevention. Particularly grievous is the bill's failure to deal with the crucial element in preventing smog—the introduction of new engines that can virtually eliminate nitrogen oxide emissions. Instead, the bill calls for controls to reduce nitrogen oxide emissions by about 30 percent.

But beyond its persistent application of the control philosophy, and the half-hearted attempt to apply prevention to automotive fuels, in one major provision the Bush bill represents a contemptuous inversion of the principle of prevention. This section, much acclaimed by certain economists, establishes a free market in pollution. Polluters will be allotted a certain amount of pollution that can be freely emitted—presumably designed to keep ambient levels within an acceptable standard. Then they may trade pollution rights, buying them when they wish to avoid installing controls, and selling them when their emissions are less than the amount allowed. This is of course a perverse parody of the "free market." In Mr. Bush's polluted marketplace, instead of goods—useful things that people want—being exchanged, "bads" that nobody wants are traded. It is a market that cannot operate unless it is provided with what it is supposed to exchange—pollutants. This is a proposal that not only fails to prevent pollution but actually *requires* it.

Mr. Bush had some noteworthy accomplices in this subversion of EPA's now seriously compromised Pollution Prevention Policy Statement, and praised them lavishly in the June 12 announcement:

> I have no pride of authorship—let me commend Project 88 [an earlier pollution market proposal by Senators Heinz and Wirth] and groups like the Environmental Defense Fund for bringing creative solutions to longstanding problems; for not only breaking the mold, but helping to build a new one.

According to the Environmental Defense Fund, this collaboration was the outcome of "dozens of meetings with White House and Environmental Protection Agency staff" and a meeting on the eve of the administration's final decision about the content of the bill between Mr. Bush and fifteen leaders of the Washington-based environmental organizations. True to their commitment to the strategy of control, the "environmental leaders" appear to have happily joined in repelling its rival—pollution prevention—at the White House gates. Their collaboration with Mr. Bush gave him a valuable political victory, confirming his self-declared characterization as the "environmental president."

In the twenty-year effort to resolve the environmental crisis, government agencies, the EPA in particular, and public organizations—some new and others newly expanded—have played a major role. The effectiveness of their approach to the problems and the success of their attempts to solve them can best be judged against the one environmental strategy that works: prevention. The EPA, after investing two decades of massive but largely futile effort in the strategy of control, has just discovered prevention—but has thus far shied away from the politically intimidating task of implementing it. The old-line, Washington-based environmental organizations are locked into the strategy of control, which generates the disputes over standards that give these organizations their arena of action. They have attempted, often successfully, to resolve these disputes through legislation, litigation, and—increasingly in recent years—through amiable negotiations with the polluters. But these efforts, however well-intentioned, have accomplished little because the controls that are supposed to implement the standards are ineffectual. In response to the failure of the strategy of control, new community-based grass-roots organizations have arisen to carry the banner of prevention to the place where the production decisions that create pollution are made: the corporations. If, as we have seen, prevention is the only effective strategy, then clearly it is the grass-roots organizations that are now at the cutting edge of the growing public movement to end the environmental crisis.

9

WHAT CAN BE DONE

WHAT CAN BE DONE to end the war between nature and man, between the ecosphere and the technosphere? There are those who believe that the war can end only with the technosphere defeated, although they differ with regard to the conditions to be imposed in the treaty of peace. Some believe in a stand-off, satisfied if human society gives up further economic growth, and with it the continued attack on the environment. Others would exact a sterner tribute, requiring that the world population—and with it the present scale of economic activity and environmental stress—be reduced. At the edge of irrationality there is the view of Earth First! that the treaty should require modern industrial society to "give way to a hunter-gatherer way of life, which is the only economy compatible with a healthy land."

Regardless of the tribute exacted by nature in such a peace treaty, it would represent a defeat—a loss of things of value—to human society, the technosphere's creator and beneficiary. This statement is, of course, my own judgment. It is based on the conviction that every human life, however degraded by poverty or enhanced by wealth, has equal value; that having more than life's necessities is better than having less; that a symphony performed in an urban concert hall has a value not supplanted by the music of a lone shepherd's pipe. If it were true that we must give up such values in order to end the present destructive assault on the ecosphere, then—according to my judgment—the war would end with a tragic defeat for human society.

I take the view that the war between the spheres is mutually destructive: nature is devastated, and human society suffers, not only because of the devastation, but also because our present, environmentally destructive production systems diminish the opportunities for economic growth, especially in developing countries. This calls for a negotiated peace that takes into account both nature's need for self-sustenance and the human need not only to maintain our present level of material welfare but to increase and disseminate it, and bring an end to poverty. Finally, in defining this task in such allegorical terms, we must remember that only one of the two belligerents—we ourselves—can perform it. I am aware that there are people ready to represent themselves as advocates for the otherwise voiceless animals, forests, fields, and seas, and even for the planet itself. Yet the fact remains that of all the living things on earth, only human beings have the capacity to consciously change what we do. If there is to be a peace with the planet, *we* must make it.

In less allegorical terms, the task is this. We must recognize that the assault on the environment cannot be effectively controlled, but must be prevented; that prevention requires the transformation of the present structure of the technosphere, bringing it into harmony with the ecosphere; that this means massively redesigning the major industrial, agricultural, energy, and transportation systems; that such a transformation of the systems of production conflicts with the short-term profit-maximizing goals that now govern investment decisions; and that, accordingly, politically suitable means must be developed that bring the public interest in long-term environmental quality to bear on these decisions. Finally, because the problem is global and deeply linked to the disparity between the development of the planet's northern and southern hemispheres, what we propose to do in the United States and other industrial nations must be compatible with the global task of closing the economic gap between the rich north and the poor south—and indeed must facilitate it.

This task, of course, is not merely formidable, but truly intimidating: in its complex interaction of technical, economic, social, and political factors; in its totally unprecedented scale; and in its impact on long-standing practices and deeply held convictions. But the time is ripe for attempting even so desperate a labor, not only because of its urgency, but also because this is a time in which even heady optimism is justified. It is justified by the sudden retreat of the threat of nuclear war and by the willingness of powerful nations to reexamine—and change—their internal distribution of both goods and political power, even if thus far chiefly limited to the Soviet Union and its allies.

As we have seen, the ecosphere is under an assault that is intolerable in its present impact and likely to end in global catastrophe if nothing is done to stop it. It is also clear that we know what actions can be taken to prevent the attack at its origin, the systems of production that have been massively developed since World War II. The solution, then, is to change these production technologies in ways that eliminate, or very greatly reduce, their generation of pollutants without hindering their ability to produce the necessary goods and services. The new technosphere must be both productive and compatible with the ecosphere. Indeed, it is evident that such a change can usually improve the *long-term* economic efficiency of the system of production. But all this is only logic, a set of sensible but abstract ideas. Present reality is both illogical and so massive and entrenched as to appear unalterable. Are we then condemned to an ongoing and ultimately suicidal war with the planet?

It should be clear, if only from the tone—let alone the content—of what I have written that I am convinced that peace is possible. A historic perspective is useful here, for it reminds us that the technology of production, the intrinsic structure of the technosphere—unlike the ecosphere—is a human creation and has been repeatedly designed and redesigned over the course of human history. From the first stone tools, through primitive agriculture and crafts, to modern mass production systems, technologies and the social structures created to implement them have come and gone. The alterations with which we are particularly concerned, which have generated the environmental crisis, have been centered in time around World War II. The profound changes in the means of production that have occurred since that landmark in the history of technology testify not only to their possible scope, but also to their speed. In less than forty years, we have managed to create a new ecological phenomenon—smog—by redesigning the automobile engine; algae-fouled lakes and seas by changing how crops are fertilized; and massive, intractable radioactive wastes by reformulating military explosives and attempting to reorganize the electric power industry. If these technological changes have been made in so short a time, surely new changes, comparable in magnitude, that undo their damage are possible as well. But the transformation will require a new motivation, for the older one—the maximization of short-term private profit—is itself responsible for the unfortunate link between the current technologies and their heavy impact on the environment. On the other hand, the possibility of such pervasive changes in the systems of production is demonstrably within the bounds of recent reality.

The practical means of carrying out these changes are evident from the twenty years' experience with the ecologically harmful production technologies, and with alternative, ecologically sound ones. In agriculture, integrated pest management is a proven method of sharply reducing the use of pesticides, and organic farming is a demonstrably successful way of eliminating the entire range of agricultural chemicals. Experience with both of these is sufficient to guide the shift from conventional agriculture to organic farming, with little or no reduction in crop yield and, indeed, economic benefits to the farmers. In the energy area, there is now a good deal of practical experience with energy conservation in residences, industry, and transportation, which only needs to be extended universally. All of the solar energy technologies that can replace gaseous, liquid, and solid fossil fuels are in hand; some are already economically competitive with conventional sources, and many are rapidly approaching that point. The most convenient source of solar electricity, photovoltaic cells, is already well developed; they are now cost-effective in areas not supplied from central sources and could soon compete with central power plants if production facilities were expanded. Several technologies that can convert cars and trucks to solar fuels—ethanol and methane from biomass—are in hand, and solar sources of electricity can be used to drive electric-powered vehicles.

Petrochemical products that are now used in a wide range of industries and consumer goods represent the most difficult problem. Substitution of current chemicals with less toxic ones is sometimes effective. Often, however, the substitutes are only a partial improvement over the present ones; thus, less chlorinated CFCs still represent a threat to the ozone layer and so-called biodegradable plastics, unlike paper, are not in fact capable of biological disposal by composting. However, it is important to remember that nearly every petrochemical product is a *substitute* for some preexisting product made of natural materials such as wood, cotton, or paper or of common materials such as metal and glass. Hence, many current petrochemical products could readily be replaced by one of these older and more ecologically sound materials. Unique, irreplaceable petrochemical products, such as pharmaceuticals or videotape, represent only a very small fraction of the total output, so that substitution could sharply reduce the output of hazardous petrochemical products and the wastes generated by manuacturing them. Recycling is sufficiently developed to convert 80 to 90 percent of the trash stream into useful materials, obviating the need for inherently hazardous incinerators. The methods for restoring natural resources such as forests and wetlands are well developed.

In sum, the technological basis for the transformation of the present systems of production to ecologically sound ones is largely in hand. However, the mere existence of the required technologies is far from enough; a great deal depends on *how* they are used. Indeed, there are already numerous examples which show that, improperly applied, ecologically appropriate technologies can do little good and sometimes more harm than good. It is possible, for example, to design a process for producing ethanol from agricultural crops that—quite unnecessarily—interferes with food production and uses more energy than it produces; such a design has been described by Mobil Oil. Similarly, photovoltaic cells have been used to construct large, centralized power stations, a procedure that destroys a major economic advantage of these devices: that they can eliminate the cost and energy losses involved in transmitting electricity.

Such mistakes can be avoided by keeping an important principle in mind: proper *integration* of the ecologically sound facility into the overall system of production. Thus, the clash between food and ethanol production can be avoided if the process is integrated into the overall system of agriculture by redesigning it to produce extra carbohydrate (see chapter 5). In the same way, photovoltaic cells become cost-effective in comparison with utilities if they are decentralized and used in local installations to drive devices, such as cooling equipment, that have a high power demand when sunlight is most available.

The transition problem must be confronted as well. Clearly, the transformation to ecologically sound systems of production will take time, and it is important to make changes within the *present* systems that will facilitate the introduction of the new technologies. The relation between natural gas and solar methane is a good example of this principle. Natural gas is a nonrenewable fossil fuel, largely composed of methane, that has several environmental advantages over oil or coal. When burned, it produces less air pollution, for example from sulfur dioxide; it yields much less carbon dioxide per unit energy produced, and therefore has a lower impact on the greenhouse effect; it can effectively fuel decentralized cogenerators in urban areas. Nevertheless, it will be necessary to replace natural gas because burning it does contribute to global warming and, as a nonrenewable fuel, it will escalate in price. Given these facts, environmental improvement can begin with the substitution, within the existing fossil fuel energy system, of natural gas for oil and coal (and nuclear power as well) in the many facilities where this is technically feasible. The investment in these methane-powered facilities and an expanded natural gas pipeline system would in turn facilitate the entry of solar methane as a replacement for natural gas. As decentralized sources of solar methane—for example, from sewage, manure, and various forms of biomass—are developed, they can be introduced into the pipeline and gradually replace the natural gas, which in this way serves as a bridging fuel. Thus, measures that initially improve the energetic and environmental efficiency with which nonrenewable fossil fuel is used can also facilitate the transition to renewable, solar energy.

It is also important to keep in mind the several aims of the transformation in production technologies, for improperly satisfying one of them may ricochet into another problem area. The transformation has three general purposes: to prevent local pollution from the systems of production; to prevent the potential worldwide environmental effects—global warming and the destruction of stratospheric ozone; to accelerate ecologically sound economic development in developing countries. If these goals are approached piecemeal, there is a danger that the method used to reach one of them will interfere with the others. The classic example is the proposal that nuclear power plants should be greatly expanded in order to replace plants that burn fossil fuels, and thereby prevent global warming. Certainly, this would reduce carbon dioxide emissions and hence the threat of the greenhouse effect, but at the expense of environmental quality: there would be a huge increase in potential radiation accidents and the still unsolved problem of long-lived radioactive waste disposal would be greatly exacerbated. In contrast, replacing fossil fuels with solar energy not only eliminates the greenhouse effect but reduces environmental pollution as well.

A similar issue arises in introducing alternative fuels to reduce vehicular air pollution. Natural gas, methanol, and ethanol have been suggested. While all of these would improve the environmental impact of gasoline and diesel fuel, methane and methanol would contribute to the greenhouse effect and methanol combustion would produce formaldehyde, a carcinogen. On the other hand, ethanol, made from current crops, is a solar fuel that would not add to carbon dioxide emissions, and is relatively free of noxious emissions. Moreover, ethanol production from biomass is a technology appropriate to Third World countries, and the development of low-cost, efficient ethanol facilities would assist them in establishing a self-sufficient energy system.

Only too frequently, remedial measures are proposed that would narrowly benefit industrialized countries, but not the developing ones that need them most. Energy conservation is an interesting example. Energy conservation—that is, improvements that reduce the amount of energy from fossil fuels needed to achieve a given productive output—is an unequivocally good thing to do. It reduces both environmental impact and economic cost. However, when elevated to the level of an *exclusive* prescription for preventing global warming worldwide, conservation of fossil fuels, while useful and necessary, is not sufficient. In this context, energy conservation has a major defect: its value as a means of reducing carbon dioxide production is gradually diminished as production and the accompanying use of energy are expanded for the sake of economic development. The point to keep in mind is that conservation of fossil fuel energy reduces carbon dioxide production, but not to zero. Hence, further economic development will eventually eliminate the initial advantage.

Thus, reliance on conservation of fossil fuels as the exclusive, or even the main, strategy for preventing global warming will tend to hinder further economic development, which is crucial in Third World countries. For example, energy conservation measures that improve the efficiency with which fossil fuels are used by 50 percent would be nullified, with respect to carbon dioxide output, when productive activity is doubled—a goal that developing countries might hope to achieve in twenty years or less. To be compatible with economic development, action to prevent global warming must include the transition to solar energy as well as energy conservation.

Is there time enough to accomplish such a massive transformation in production technology? We know enough about most of the environmental problems to answer this question, at least approximately. Current computer models indicate that the most potentially catastrophic problem—global warming—could be prevented by appropriate actions over the next few decades, but beginning now. Indeed, for reasons quite apart from global warming, it will be necessary in the next twenty years or so to accomplish the main change needed to prevent the greenhouse effect: replacing the present massive use of carbon-dioxide–generating fossil fuels with solar energy. As pointed out in chapter 5, the reason is economic rather than ecological, and derives from a basic, economic fault inherent in the use of nonrenewable fossil fuels: the cost of extracting them rises exponentially with time, demanding a progressively greater fraction of the output of the economic system to produce a given amount of energy. This becomes, in time, an intolerable drain on the economic system.

The ozone depletion problem appears to be intensifying on a similar time scale, suggesting that the effort to replace CFCs and other industrial products that destroy ozone, which has just begun, must succeed in this aim within the next decade or two. Because of the time delays inherent in the processes that generate both global warming and the ozone effect, the campaigns to combat them must begin immediately. While the remaining, more local environmental problems may not generate catastrophes of global dimensions, they are, of course, sufficiently serious to demand remedial action over an even shorter time span, perhaps ten years or so rather than twenty to thirty years.

What will it cost to transform our present systems of production into ecologically sound ones? Stated in economic terms, resolving the environmental crisis is a matter of investment policy. If, as we have seen, the only effective way of improving environmental quality is to appropriately redesign the technological systems of production, then the economic task is to invest in the necessary new production facilities so that polluting ones can be closed down. Two subsidiary issues arise: Is sufficient capital available to make the new investments over the necessary time span? What motivation, or principles of governance, will ensure that the investments are made in the proper kinds of productive technologies? The answer to the first question can be derived from the time scale and other considerations. (The second one is discussed in the following chapter.)

The effective life span of a large-scale capital investment such as a power plant is about twenty to thirty years, so that if sufficient capital is available from the existing plants' operation, the entire complement of existing power plants will ordinarily be replaced over that time span. This suggests that the anticipated rise in global temperatures (or at least that part of it due to carbon dioxide generated by power plants) could be prevented by beginning now in industrialized countries to replace each obsolete fossil-fueled plant with a comparable source of renewable, solar energy. (For reasons already described, I exclude nuclear power from consideration.) Of course, energy conservation is an essential accompaniment to this replacement process, but as indicated earlier, is not itself sufficient to resolve the overall, global problem. Energy conservation measures are so cost-effective that the initial investment is recovered, through savings, in a relatively short time, generally two to five years. Hence, there is no economic barrier to this process. However, especially in residences, the necessary capital is often unavailable (for example, to householders of modest means), so that some form of subsidization is needed. The present federally funded weatherization program is an example; but its funding level is so low that only 10 percent of the eligible (low-income) households and a far lower percentage of all households have been weatherized. Thus, the issue is to redirect investment policy in the power system, with the needed capital provided by the investment funds normally used to replace obsolete facilities. Some additional funds will be needed, however, to increase the present rate of energy conservation. A Worldwatch Institute estimate of the cost of energy conservation generally is $33 billion per year over a ten-year period.

Similar considerations apply to other industries, such as automobiles, in which capital funds regularly available from the industry's earnings would be invested in new facilities to produce new, more environmentally sound vehicles. However, additional capital will be needed to modernize and electrify the U.S. railroad system as a means of reducing fuel consumption by trucks and intercity automobile traffic. An estimate by Dr. J. I. Ullman of Hofstra University suggests that in the United States this might amount to expenditures of $10 billion per year over a ten-year period.

In the petrochemical industry, matters are much more complex. In many instances, it will be necessary to do more than merely change the technology of petrochemical production—for example, by using more of the organic chemicals that occur in nature, such as ethanol. It will also be necessary to replace petrochemical products, especially plastics, by increasing the facilities that produce substitutes such as metals and glass. Investment capital will be needed to build such production facilities faster than the present rate of replacing obsolete plants. If, for the sake of argument, most of the present petrochemical facilities were gradually phased out and replaced by new plants to produce the more ecologically sound substitutes, the new capital needed might be very roughly approximated by the size of the fixed capital now in the petrochemical industry—about $200 billion. Thus, replacing the bulk of the petrochemical industry with alternative, ecologically sound production processes might require an investment of about $20 billion annually for the next ten years. Clearly, this would need to come from sources outside the petrochemical industry itself.

The same approach can be used to estimate the cost of replacing the present systems of producing fossil fuels, a transformation that might take place over a twenty-year period. The total capital goods now invested in coal mining, oil and natural gas production, and petroleum refining is about $500 billion. If we assume that a similar level of outside capital investment would be needed to create the necessary solar energy facilities, the required cost is about $25 billion per year over the twenty-year period.

In agriculture there is a special reinvestment problem that relates to the condition of the soil. The heavy use of agricultural chemicals is generally accompanied by the continued loss of natural fertility and of the natural insect-controlling elements, such as predators. U.S. experience suggests that a minimum of about five years of nonchemical agriculture—and on the average somewhat longer—is required to restore the natural systems of fertility and insect control. In the intervening period, income may be significantly below that achieved by conventional chemical-based farming. In that period, farmers may need subsidies to carry them through the transition. It is difficult to know how this external support would compare with the level of subsidization provided to U.S. farmers in recent years, which has averaged about $8 billion per year (rising to $18 billion in 1987), but the amount may well need to be larger, perhaps $12 billion per year.

It would appear, therefore, that the transformation of the major systems of production that are required to restore environmental quality could be carried out in an industrialized country such as the United States in part by redirecting the investment of the capital generally needed to replace obsolescent production facilities at present. Additional capital amounting to about $100 billion per year over a ten-year period would be needed as well to install energy conservation measures, to modernize and expand the rail system, to support replacement of a good deal of the petrochemical industry, to replace the present energy system with a solar one, and to subsidize the agricultural transition. In this connection, it is interesting to note that the United States has spent about $70 billion annually over the last ten years in the largely futile effort to improve the environment by attacking the symptoms rather than their origins. The issue of further expenditures needed to accomplish the technological transition on a global scale, including support for economic development in Third World countries, is taken up in the following chapter.

What institutional mechanisms would facilitate introducing the social interest in environmental quality into the decisions that govern production? In the United States, certain provisions of our regulatory laws at least imply that environmental improvement may be achieved by the socially mandated choice of production technology. Thus, the basic instrument created by the National Environmental Policy Act, the environmental impact statement, according to Section 102C of the act, must include a presentation of "alternatives to the proposed action" in addition to a description of unavoidable adverse environmental effects of implementing the proposal. This provision implies that in, say, a public hearing on a trash-burning incinerator that unavoidably emits dioxin into the environment, citizens could call for a decision mandating the choice of a technology that would separate and recycle trash rather than burn it. The banning of the use of DDT in agriculture or of PCB in manufacturing is a concrete, if relatively narrow, example of this approach. Opportunities, then, do exist within the framework of present environmental legislation which, though they have been little used thus far, would permit at least the public weighing of alternative choices of production technologies.

Now, with pollution prevention adopted as EPA policy—if not yet practice—such unused legislative vehicles may be brought out of storage. On the other hand, given the overwhelming dominance of control strategies in both the legislation and its implementation, it is evident that an entirely new legislative base will be needed to redirect the environmental program from a strategy of control to prevention. It will be necessary, I believe, to go back to the purpose of NEPA—"to prevent and eliminate damage to the environment and the biosphere"—and translate that unequivocal intention into new legislation that specifically implements that goal.

Another existing but little used procedure that might facilitate the transformation of production technologies is the market power of government purchases. The growth of one of the country's great entrepreneurial successes—the computer industry—was in fact accelerated in this way. In the 1950s, the U.S. military establishment, not to speak of the political one, was convinced that the nation had to prepare for nuclear war by developing missiles capable of minute and intricate maneuvers to find their Soviet targets. Electronic computation devices that could perform this task existed, but they were far too large to be tucked away in a missile's nose cone. The answer was a new device—the integrated circuit computer chip—which enormously compressed the space needed for complex electronic systems. At the time, the United States had no large-scale facilities for producing such devices, which would be needed in the many thousands to supply the new missiles. Undeterred, the Pentagon used a simple economic device to persuade private entrepreneurs in the fledgling semiconductor industry to undertake the task: they offered them large and lucrative contracts. Armed with a contract, an entrepreneur could borrow enough money to set up highly rationalized production facilities capable of output on the scale that the Pentagon demanded.

This created an industry capable of manufacturing a large number of integrated circuit chips. Since many missiles—and therefore computer chips—were ordered, the production facilities expanded and improved their efficiency with extraordinary speed. In a few years, a computer chip that originally cost $50 came down in price to $2.50. Suddenly, it was possible to build pocket calculators that sold for a few dollars. Not much later, we had the word processors and computers that have not only become a major industry but have profoundly changed the way in which work everywhere is done.

Some twenty-five years later, during the Carter administration, an effort was made to resurrect this practice, this time for the development of the photovoltaic cell industry. Photovoltaic cells, which convert sunlight directly into electricity, like computer chips, are semiconductor devices. In the 1970s, photovoltaic cells were produced from large crystals of silicon that were sawed into very thin slices—a difficult and inefficient process. As a result, the cells were very costly. The price then was about $20 per peak watt (a peak watt represents the amount of DC power a cell will produce when the sun is at its peak intensity). At that price, photovoltaic cells were economically feasible only for a few special applications, such as the operation of remote weather stations.

At the time, the total manufacturing capacity of the U.S. photovoltaic cell industry was less than one megawatt per year. An analysis of the "learning curve" that reduced the price of computer chips showed that a 150-megawatt order would bring the price per peak watt of photovoltaic cells down from about $20 to $2-to-$3 within a year, to $1 in three years, and to 50¢ in five years. As the price drops, the market expands enormously, so that, for example, at $1 per watt photovoltaic cells would become economically feasible for roadside lighting, representing a demand of about 50,000 megawatts. At 50¢ per peak watt, photovoltaic cells were expected to be competitive with electric utilities in many parts of the country. Congress was persuaded that this would be a sensible bargain, and a bill for the purchase of nearly half a billion dollars' worth of photovoltaic cells for federal installations was passed. The bargain was never realized, for the bill was vetoed by Mr. Carter.

This same strategy of federal purchases could be used to improve environmental quality by producing a smog-free engine. According to a National Science Foundation study, such an engine—the stratified charge engine—existed in 1974 as a prototype model in the hands of the Ford Motor Company. The engine was capable of achieving the 90 percent reduction in nitrogen oxide emissions mandated by the Clean Air Act and could therefore sharply reduce smog production. The report showed that if a decision to build the engine had been made in 1974, by 1988 it would have replaced the smog-generating engines in the entire fleet of U.S. passenger vehicles.

The federal government buys more than $7 billion worth of cars and trucks annually, based on highly detailed specifications. Suppose that the specification for a stratified charge engine, or some other design capable of qualitatively reducing emissions, was added to the bid requirements. Surely one of the Detroit companies, or possibly a Japanese competitor, would respond and grab the huge order. Once that happened, the total replacement of the present smog-generating engines would be inevitable, if only because public demand is likely to insist on the benefits of smog-free engines being spread from the federal government to the taxpayers who support it.

The same could be done in many other fields. For example, now that we know that a system of intensive recycling can replace trash-burning incinerators, the federal government could readily require that federal installations dispose of their trash through such a recycling system. This would immediately create a large demand for the required materials recovery facilities and compost plants, converting a now minuscule industry into one sufficiently large to come knocking on the doors of the country's cities and towns for their trash disposal business. In the same way, government purchase requirements that favor recycled materials—paper, for example—would lead quickly to the expansion of recycling facilities.

Thus, government purchasing power is a force—little recognized but occasionally used with considerable effect—that could influence investment decisions and implement the changes in productive technologies needed to eliminate a number of pollution problems.

Another force capable of influencing production decisions, at least in certain areas, is the consumer. In 1989, Alar, the treatment for enhancing the marketability of apples, provided a particularly instructive example. Like many other petrochemical products, Alar represents a health risk, for it induces cancer in test animals. As in many other cases, there have been controversies about the resultant hazard to people, especially children, and about what standards should be applied to limit exposure to "acceptable" levels. Following a well-publicized Natural Resources Defense Council report that some apples contained enough Alar to exceed the cancer risk guidelines for children—dramatized by an appearance of the actress Meryl Streep before a congressional hearing—there was a sharp drop in the sales of apples and apple juice. Apple growers vainly asserted that the risk was "small" and with the debate intensifying, mothers became increasingly apprehensive.

This was the customary, futile pattern of environmental debate. But Alar broke out of this pattern when the manufacturer, Uniroyal, decided—apparently under the pressure of reduced apple sales—that regardless of the toxicological and regulatory uncertainties, Alar would be taken off the market. This in turn reflected an equally simple fact: parents were unhappy about raising their children on apple juice that represented *any* threat to their health; food, after all, is supposed to be good for you. The Alar story illustrates the role that public opinion can play in preventing environmental hazards. Parents were not inclined to argue about how much Alar was tolerable; they wanted *none* of it in apples, and Uniroyal responded by an action to ensure exactly that.

Another consumer campaign against petrochemical products has recently begun to take shape, with throwaway plastics as the main target. Beginning with a law banning certain plastic throwaway items in Suffolk County, Long Island, similar measures have been adopted or are under consideration in cities as large as Minneapolis. While the plastics industry has successfully resisted these laws, partly by hurriedly establishing recycling centers for Styrofoam throwaways, public opposition appears to be gaining strength. Jeffery Hollender (who has been described by a journalist as a "one-man seismograph recording the jolts and tremors of fads and vogues") has recently proposed lists of do's and don'ts for the waste-conscious consumer. Nearly every item in a long list of don'ts is made of plastic or some other petrochemical product. The do's are products made of materials such as soap, paper, wood, glass, and metal that the petrochemical industry has been pushing out of the market. If this potential vogue catches on, the petrochemical industry will face public opposition to their production processes in a painfully meaningful form—sales and profits.

It would appear, therefore, that ecologically sound replacements for the current, highly polluting production technologies exist and that there are at least some mechanisms that could exert pressure on private entrepreneurs to make the necessary investments. Will this be sufficient to resolve the environmental crisis? Can the ideological barrier that guards private investment from social intervention be skirted in this way—or will we need to resolve the ideological issue itself?

10

MAKING PEACE WITH THE PLANET

THE INCREASINGLY ACUTE global environmental crisis can only be resolved by a comprehensive transformation of the present systems of production—in agriculture, manufacturing, power production, and transportation. The preceding chapter shows that despite its intimidating size and complexity, this task *can* be accomplished. The required ecologically sound production technologies exist today and could be introduced in time to avoid the impending global disasters. They would prevent global warming and ozone depletion, effectively resolve most of the world's pollution problems—such as smog, acid rain, and toxic chemicals—and reduce the mounting accumulation of trash. Once in place, the new systems of production, in the long term, are likely to be more productive economically than the present ones. Ecologically sound production technologies would enable Third World countries to develop without enduring environmental degradation in exchange. In sum, the material requirements for ending the disastrous assault on the ecosphere are in hand.

Moreover, the need for undertaking the task is widely recognized. There is now a worldwide awareness that the environmental crisis must be solved. Public pressure has become strong enough to persuade political figures not previously known for their ecological concern—President Bush, Prime Minister Thatcher, and President Mitterrand are outstanding examples—to declare a newfound environmentalism. Speaking to the United Nations, President Gorbachev has pledged the Soviet Union to work equally for world peace and world environmental quality. In Europe a "Green Wave" has carried the environmental issue to the top of the political agenda, reinforced by the election of Green Party candidates to most European parliaments. In 1990 the twentieth anniversary of Earth Day 1970 promises to bring these forces together and lift environmental concern to a new peak. The environmental crisis is no longer the environmentalists' issue; it has been adopted, globally, by the political mainstream. In sum, both the desire to resolve the environmental crisis and the means of doing so exist. Are we, then, on the brink of actually beginning this monumental task?

History warns against overoptimism. In April 1970, at least in the United States, we also appeared to be on the edge of success. The spontaneous public outcry over the state of the environment led quickly to promises of action from almost every political leader, beginning with President Nixon. Laws were passed and an elaborate environmental regulatory system was established. Many billions of dollars have been spent. Yet, as we have seen, this vast effort has failed. It failed because the effort was misdirected, dealing only with symptoms and applying palliatives instead of attacking the problem at its root.

Now, as we reach the twentieth anniversary of the monumental failure to create a workable environmental program, there is a danger that history will repeat itself. William K. Reilly, the new administrator of the EPA, has already been fitted with a pair of analytical blinders, and bureaucratic sophistry promises to negate the agency's recent discovery that only prevention can deal with pollutants successfully, at their origin. In an article in a recent issue of the official *EPA Journal,* fashionably titled "The Greening of EPA," Mr. Reilly concludes that "the time has come to consider applying market incentives/pollution prevention approaches to environmental programs across the board." By "market incentives," Mr. Reilly means the provisions in the administration's Clean Air bill that encourage corporations to participate in a market in which the right to pollute the environment is bought and sold. He apparently believes that trading pollutants and preventing them are so harmonious that they can be joined, typographically, with a slash. But of course the two ideas are totally incompatible: prevention means *zero* emissions, so that there are none to trade. But there is an even deeper, ideological contradiction between pollution prevention and the free market: pollution prevention means governing the design of production processes in keeping with the *social* interest in environmental quality. In the United States, the free market operates on the principle that production processes are under *private* governance and are therefore designed solely as profit-maximizing responses to market forces.

Mr. Reilly's self-contradictory position exemplifies the basic fault in the present system of environmental regulation: the root of its failure lies in the realm of reasoning. Little or no effort has been made to ask *why* we are in the grip of the environmental crisis. When a variety of alternative explanations—overpopulation, affluence, or ecologically unsound production technology—were offered, they were simply noted and, in designing remedial action, ignored. The answer (control the symptoms) was given before the question (what causes them?) was asked. And now that the evidence of failure has forced the EPA to consider prevention—the remedy that attacks pollution at its source—a forced marriage with its opposite, a market in pollution, threatens to nullify this promising innovation.

Yet logic leads us directly to the basic, unasked question that lies at the root of the environmental crisis. As we have seen, there is a broad consensus that it is in the national and global interest to restore the quality of the environment, and the resultant legislation confirms that accomplishing this aim is a social and hence governmental responsibility. As we have also seen, meaningful environmental improvement requires the proper choice of technologies and systems of production, so that this choice becomes, in turn, a social responsibility. But in our free-enterprise economy, the right to make this choice is in private, not public, hands—a principle that very few Americans will even discuss, let alone challenge. The basic question, then, is this: To what extent should the choice of production technologies be governed by private, generally short-term economic considerations such as profit maximization, and to what extent by long-term social concerns such as environmental quality?

The recent entrepreneurial history of the United States Steel Corporation (now renamed USX) illustrates the problem. In the 1970s, the company developed a plan to build a large, modern low-pollution steel mill at Conneaut Lake, Pennsylvania—a socially useful way to reduce both steel imports and pollution. Then in 1982, with the price of steel driven down by cheap imports and the price of oil rising, U.S. Steel decided to abandon the Conneaut Lake project and instead bought the Marathon Oil Company. By 1985, 54 percent of the company's business was in oil, and only

35 percent in steel. As the economist Robert Reno commented:

> Now, it is arguable whether the best interest of the nation would have been better served if U.S. Steel had reinvested in modern steel plants rather than purchasing Marathon Oil. Anyway, there's no law that says corporations must choose national interests over those of their stockholders.

In fact, the company's legal obligation is not to the nation but to its own stockholders, and that obligation is not to produce steel but short-term profits. This business maxim has been crisply stated by John Swearingen, who, while president of the Standard Oil Company of Indiana, had promoted a shift toward chemical production: "We are not in the business of producing energy, but the best possible return on the stockholders' investment."

This is, after all, the meaning of "free enterprise": the owner of capital is at liberty to invest it in whatever enterprise offers the most promising rate of return, market share, or some other private advantage, whether it produces steel, chemicals, oil, or plastic swizzle sticks. And, as we have seen, this right has been exercised regardless of its environmental consequences.

The institution that largely determines the course of production technology in the United States—the corporation—is itself a peculiar amalgam of social and private features. In one sense, a basic purpose of the corporate form is social: to accumulate, from a large number of people, an amount of capital far greater than that available from any one person. Indeed, almost all the earliest American corporations were chartered for a specific social purpose—building a canal, bridge, or turnpike—that required large, collective capitalization. Today, corporations may be chartered for any legal purpose, and thus have the right to determine privately how their socially collected capital will be invested. When stockholders buy a share in the corporation, they turn over their rights of governance to the managing board in return for their single reward: expected profits. Nevertheless, corporations are certainly still social in their effects—not only on the environment, but on employment, wages, working conditions, and the fate of whole communities as well. Yet with respect to their control, corporations are private, for they are governed by a small group of self-perpetuating officers who have no general accountability to society other than adherence to relevant government regulations, or—apart from profitability—even to the stockholders who own the corporation.

The conflict between the modern corporation's social role and its private control led Adolph A. Berle and Gardiner Means to declare, in their classic work *The Modern Corporation and Private Property*:

> The economic power in the hand of the few persons who control a giant corporation is a tremendous force which can harm or benefit a multitude of individuals, affect whole districts, shift the current of trade, bring ruin to one community and prosperity to another. The organizations which they control have passed far beyond the realm of private enterprise—they have become more nearly social institutions.

From their detailed analysis of this conflict, Berle and Means concluded that the separation of corporate ownership from control places society "in a position to demand that the modern corporation serve not only the owners, but all society. . . . It remains only for the claims of the community to be put forward with clarity and force."

The date of Berle and Means's work, 1932, is significant; it suggests that the shattering impact of the Great Depression on confidence in the free-enterprise system—capitalism—had impelled them to seek alternatives. Since then, other potential critics, if not silenced, have at least been inhibited by the remarkable vitality of the U.S. economy after World War II, the dazzling new products and the growth of the national and private wealth. Now that the attendant social ills—persistent unemployment, insufficient affordable housing, inadequate medical care, pollution—are more evident, the inhibition has been broken. For example, the 1986 Catholic bishops' pastoral letter on economics, citing Thomas Aquinas, asserts: "No one can ever own capital resources absolutely or control their use without regard for others and society as a whole. . . . Short-term profits reaped at the cost of depletion of natural resources or the pollution of the environment violates this trust."

The bishops' letter reflects the 1981 encyclical "On Human Work," in which Pope John Paul II asserted, on moral grounds, that ownership does not justify exclusive control of production facilities and declared that "in consideration of human labor and of common access to the goods meant for man, one cannot exclude the socialization, in suitable conditions, of certain means of production."

One can, of course, embellish these observations and arguments with a doctrinal term—socialism (classically defined as social ownership and control of the means of production)—and embrace or dismiss them by reacting to that term. While this may satisfy one's ideological convictions, neither approval nor disapproval can alter the reality, which, as we have seen, is that substantial environmental improvement can occur only when the choice of production technology is open to social intervention. If the national commitment to environmental improvement is to be honored, we must respect this reality and find suitable ways to implement the social governance of production.

That American society is not yet prepared to undertake this task is evident from the shocking public response to the Catholic bishops' pastoral letter and to the pope's encyclical before it—nearly total silence. The United States, after all, is blessed with a literate population: schools and universities that teach economics and even the precepts of public morality; ravenous news media that feed on controversy; and politicians frequently called upon to pass judgment on both economic and environmental issues. Surely so profound a challenge to conventional wisdom from such a responsible source should have instigated a lively debate within and among these sectors of society. Yet, apart from complaints by a few right-wing ideologues who happened to be Catholics and resented this ecclesiastical challenge to their business practices, the bishops' letter quickly disappeared from public sight. In the *New York Times,* for example, we find, on the day of publication, excerpts from the letter and a commentary on it by the economics columnist, Leonard Silk. Four days later an op-ed piece by Patrick Buchanan attacked the letter. Since then, silence.

The silence that greeted the bishops' pastoral letter is one of many expressions of a deep-seated conviction in the United States that our economic system automatically selects those productive enterprises that use natural, economic, and human resources most efficiently and that therefore best serve society. Although the virtues of the system are nearly universally acclaimed, it is rarely called by its name: capitalism. This is regarded as a breach of political etiquette, introducing partisan ideology into what is supposed to be an unspoken national consensus. Only brash ideologues like Malcolm Forbes assert that they are, in fact—like their more silent colleagues—capitalists. Most capitalists prefer to be described as "businessmen" or "industrialists," neutral terms not subject to an ideological challenge.

But if we wish to resolve the environmental crisis—not to speak of the other social issues that suffer from the grip of private governance on most of the nation's resources—the ideological issue can no longer be evaded. It may be argued that market mechanisms, such as those discussed in the preceding chapter, which are available to us now with no ideologically driven change in U.S. capitalism, will do the trick. But these mechanisms—federal purchases to create a market for new ecologically sound products, or consumer pressure against unsound ones—have a serious limitation: the market, and hence the undesirable product to be displaced, must already exist. Thus, federal purchases might create a market for a smog-free car, but only because of public dissatisfaction with the existing, smog-generating ones. Similarly, the small but rapidly growing market in organic fruits and vegetables has been created largely because of public dismay over the existing pesticide-laden products. As an expedient way of dealing with the present array of products generated by past ecologically unsound production decisions, these market mechanisms make a certain amount of sense. But they could work only because we have already gone through the wasteful, destructive experience of fouling the air with smog and food with pesticides. They will not protect us from new mistakes. The evolution of technology will continue, and left to their own short-term profit-maximizing devices, future capitalists will make new environmental mistakes. The prospect of enduring them for a generation or so before the ecological faults are detected and the production processes responsible for them are gradually defeated in the marketplace is hardly a comforting answer to the ongoing war between the ecosphere and the technosphere. Here, too, on the level of the decisions that govern production technology, only prevention makes sense.

There is, of course, a ready response to the conclusion that both the origin of the environmental crisis and our failure to resolve it can be traced back to the capitalist precept that the choice of production technology is to be governed solely by private interest in profit maximization or market share. If that is true, what can account for the bad environmental record of the Soviet Union and other socialist countries? By far the worst cases of radioactive contamination, of which the accident at Chernobyl is only the most spectacular, have occurred in the Soviet Union; Czechoslovakia and Poland have the highest levels of industrial pollutants in Europe, and perhaps the world; the socialist countries, especially the Soviet Union, have rapidly adopted the same agricultural chemicals that are responsible for polluted water supplies and pesticide-contaminated food in the United States and Europe. If private governance of production—the characteristic feature of capitalism—is at fault, why do the environmental problems that it generates also occur in socialist countries where production decisions are presumably under social or governmental control?

One reason is that—as the Soviet government now admits—until recently, advocates of social interests such as environmentalism have not been free to comment on, let alone influence, government decisions. A less obvious—but decisive—reason is that most of the the systems of production that the socialist countries have adopted were in fact developed in the capitalist countries after World War II: for example, chemical agriculture, nuclear power plants, and the petrochemical industry. Having been developed with no concern for their environmental impact, these production systems wreak their havoc on the environment equally in capitalist countries and the socialist ones. It is, after all, unreasonable to expect that the automobiles produced in the Soviet Union at Togliattigrad, by a plant imported intact from the Italian company Fiat, would refrain from emitting the same pollutants in Moscow that they produce in Rome—perhaps out of respect for socialism. It is an unassailable if ironic fact that the economic development of the Soviet Union and its socialist neighbors after World War II has been based on the major new production technologies developed in the United States and other capitalist countries, where their design was guided by a capitalist motive: short-term profit maximization—to the exclusion of environmental and other social concerns.

Some will object that technology is value-free, and that technological decisions are outside the realm of public or private policy, determined instead by "objective" scientific and technical considerations. But the evidence contradicts this view. Certainly, it is an objective scientific fact that high-compression automobile engines are particularly powerful and generate smog, and that low-compression engines are less powerful but smog-free. The decision to manufacture one engine or the other, however, is not based on scientific necessity but on human choice. Such a choice is a matter of policy, not science. The U.S. automobile manufacturers' decision to build large cars that required powerful high-compression engines was a policy judgment based on profit maximization ("Minicars make miniprofits"). In effect, the engine is not a value-free piece of technology, but an instrument of economic policy, designed to carry out a corporate purpose, with an unintended but nevertheless disastrous result: the smog that blankets every major American city.

The petrochemical industry is a particularly interesting example of how the basic features of the economic system in which it originated—capitalism—are heavily imprinted on its technological structure. As Lord Beeching, one of the industry's leaders, has said about the petrochemical industry: "Instead of producing products to satisfy existing industrial needs it is, increasingly, producing new forms of matter which not only replace the materials used by existing industries but which . . . forces existing industries to adapt themselves to use its products." In effect, the very technical design of the petrochemical industry was created to seize new markets and to maximize profit, regardless of its effect on social concerns, such as environmental quality, or the stability of other industries.

It is evident, therefore, that although the socialist precepts that are the ideological foundation of the Soviet economy should certainly include social governance of production, in practice this principle has had no discernible influence on the Soviet production system. With *glasnost,* this is changing. Environmentalists are now able to generate public discussion about environmental issues in the Soviet Union. As a result, public opposition has already forced the abandonment of a series of major projects, including the construction of nuclear power plants and ecologically hazardous hydrological projects. But despite these recent events, there is no good evidence as yet that, *as a matter of principle,* public concern with social issues such as environmental quality will govern major decisions about production technology in the Soviet Union.

In sum, neither the Soviet Union nor any other country as yet provides an institutional example of social governance of technological production decisions. Every country, whether industrialized or developing, whether capitalist or nominally socialist, uses the same smog-generating cars, noxious synthetic chemicals, hazardous nuclear power plants, and ecologically unsound chemical agriculture—except where isolated public campaigns have blocked them. Every country, to one degree or another, has reaped the short-term economic benefits—and the environmental hazards. The fact that the hazards are inseparable from the benefits is inherent in the design of the production technologies and reflects the common motivation of their capitalist creators: their exclusive interest in maximizing economic return.

Thus, every country in the world is faced with the same task of technological reconstruction and with the necessity of creating an effective means of enabling the social concern with environmental quality to govern it. In theory, one would expect that since their guiding ideology includes social governance of production, socialist countries such as the Soviet Union should achieve this goal more readily than capitalist countries. In practice, the realization of this goal will be determined by the economic and political changes in the Soviet Union and its neighbors since Mikhail Gorbachev has come to power. Under President Gorbachev's guidance, the Soviet Union is moving toward a market economy, in which supply and demand are free to determine selling price, although currently still subject to a variety of constraints. Does this mean that the Soviet Union will allow market forces—to the exclusion of social interest in environmental quality—to determine production technology?

In this connection, it is useful to examine two roles that the market can play. Once goods are produced and offered for sale, the market is clearly an effective way to facilitate their distribution, by favoring those that are desirable because of quality and/or price. While such markets have been characteristic of capitalism, there appears to be no reason why they cannot operate in socialist economies as well. However, the market in goods can also play another role that links it to the technology of production: it can dictate *how* goods are produced in order to facilitate their low price and hence their marketability. In this role, the market, immune from social concerns such as environmental quality, will tend to induce production decisions that are likely to conflict with them. In the U.S. capitalist economy, market-motivated private production decisions, as we have seen, *have* conflicted with the social interest in environmental quality. In a genuine socialist economy, where production decisions—in keeping with that ideology—are supposed to be under social governance and therefore could include environmental factors, this conflict need not arise.

Thus, the market is a useful means of facilitating the flow of goods from producer to consumer; but it becomes a social evil when it is allowed to govern the technology of production. As the economist Manfred Bienefeld reminds us, "The concept of the self-regulating market is a dangerous and socially corrosive myth . . . the market is a wonderful servant but a disastrous master." It remains to be seen whether in the Soviet Union *perestroika* can distinguish between the two roles of the market and make good on Mr. Gorbachev's promise to improve the flow of goods *and* enlist in the battle for environmental quality. The issue in the Soviet Union is whether the dynamics of *perestroika* will carry the influence of the market into the productive process, or whether the force of socialist ideology (however weakened at present), the growing political role of public opinion, and the increasing impact of labor—laborers often being the earliest victims of environmental hazards—may prevail and stop the free market at the factory gate. The very vigor of the process and the growing willingness of Soviet society to reexamine its history—regarding both theory and practice—suggests that the issue will at least be confronted there, which is, after all, the first step in resolving it in favor of environmental survival.

The response of the Bush administration to *glasnost* and *perestroika* has been one of more or less quiet satisfaction that, as they see it, the Soviet Union and its allies have begun to acknowledge the universal virtue of the precepts of the American economic system—capitalism. On the more ideological sidelines, there is glee and smugness, the latter carried to a kind of cosmic extreme by one commentator who claims that the sweeping changes in the socialist countries is a sign "of the endpoint of mankind's ideological evolution and the universalization of Western liberal democracy [read: capitalism] as the final form of human government."

It is useful, in this connection, to reflect on the reason for the unexpected emergence of *glasnost* and *perestroika* in the Soviet Union. Clearly, the driving force behind this political upheaval has been the clash between the realm of reality and the realm of ideas. For some seventy years, the Soviet people were inculcated with a set of ideas that go back to Karl Marx, as interpreted by Lenin and brutally misapplied by Stalin. Now that the reality of the country's economic decline and its people's apathetic response have clashed with the ideological expectations, the hitherto unexamined relationship between the ideology and its practical interpretation is subject to comment, criticism, and change.

There is also a clash between ideology and reality in the United States—as yet, as it was in the Soviet Union before Gorbachev, unacknowledged. As we have seen, our reigning ideology, capitalism, clashes with the reality of the environmental crisis—not to speak of the reality of our country's shameful levels of poverty, and our inadequacies in the areas of housing, medical care, and education. We in America have as much reason as the Soviet Union to engage in a *perestroika* of our own—to open to public discussion the serious conflict between our unexamined capitalist ideology and the failed effort to resolve the environmental crisis—as a prelude to radical (in the sense of getting at the root of the problem) remedial action.

In a sense, the effort has already begun, but in our typical fashion, answers are being offered long before the question is asked and properly defined. One answer is the notion of "corporate responsibility," in its most recent form the "Valdez Principles" introduced by a coalition of environmental organizations and investment groups. Taking their name from the Exxon tanker involved in the catastrophic oil spill in Alaska, the Principles ask corporations to "publicly affirm [their] belief that corporations and their stockholders have a direct responsibility for the environment . . . and [to] seek profits only in a manner that leaves the Earth healthy and safe."

This precept conflicts with the conventional view of a corporation's responsibility for profit *maximization,* as expressed, for example, by John Swearingen. It implies that the corporation will sacrifice some of its profits, if necessary, in order to leave "the Earth healthy and safe." In this sense, the Valdez Principles provide an answer—a voluntary corporate pledge to conform to social interest when it conflicts with their private interest—to the basic question of whether social governance of corporate production decisions is desirable, but does so without explicitly framing the question. If the question were clearly stated, it would be evident that the Valdez Principles fly in the face of the realities of American capitalism. Consider, for example, the impact of the Principles on the petrochemical industry, which sells numerous products—pesticides, fertilizer, and plastics, for example—that are nearly impossible to produce and/or use "in a manner that leaves the Earth healthy and safe." As noted earlier, if the petrochemical industry were required to destroy its toxic wastes instead of releasing them into the environment—surely the proper response to the Valdez Principles—the cost would be much larger than the industry's annual profit. If, under some undeniable moral compulsion, the petrochemical companies were to offer to abide by the Principles, they would be committing corporate suicide.

In effect, the Valdez Principles, if really implemented, would call upon the country's economic rulers—corporate executives who possess enormous personal wealth and the political power that it endows—to voluntarily relinquish a good deal of both. There is, to be sure, a considerable value in confronting the corporate managers with this demand and threatening them with stockholders' and consumers' reprisals if they reject it. But realism suggests that although this process may be a valuable means of stimulating public discussion, it is unlikely by itself to generate the desired changes in corporate behavior.

It seems evident, therefore, that the responsibility for introducing social concerns such as environmental quality into corporate production decisions lies with society itself. As Berle and Means have pointed out, society is "in a position to demand that the modern corporation serve not only the owners, but all society." To achieve that purpose, we need to follow their further admonition that the "claims of the community . . . be put forward with clarity and force," rather than relying on the corporation's ecological goodwill. In sum, institutional means must be found to implement the social governance of production decisions—a power essential to resolving the environmental crisis, and very likely other social issues as well.

A critic will respond to this conclusion by pointing out that it means little to call for "social governance" of production decisions without specifying *how* this might be accomplished. After all, such an intercession has never been attempted in the United States and other capitalist countries, and it has failed to materialize even in the favorable—in theory at least—climate of socialism. It will be pointed out, moreover, that a social mandate for ecologically sound production decisions would need to be implemented through some form of planning—which has clearly been unable successfully to govern the performance of the Soviet Union's system of production. Given this record, the view that social governance of production—environmental democracy—could ever become reality takes on the cast of illusion.

Yet recent history admonishes us to expect the unexpected from the inherent impulse to extend democracy. In just two years, 1988 and 1989, the world has witnessed a series of startling outbreaks of popular democracy—powerful and in most cases amazingly successful movements that have arisen, spontaneously, outside the framework of conventional politics. In East Germany, in two remarkable months masses of people found ways to express the ingrained sense that freedom of movement is essential to the human condition—and brought down the Berlin Wall, transforming a seemingly unshakable regime and remaking the history of Europe. In Czechoslovakia, in less than two weeks, repeated massive rallies forced the repressive government installed by the Soviet invasion in 1968 into a coalition with the same dissidents it had previously jailed. In an equally short time, Rumania deposed its corrupt and autocratic ruler and instituted democratic reforms, unfortunately at the cost of considerable bloodshed. In Poland the innate democratic impulse took the form of an autonomous labor union, Solidarity, that despite its suppression under martial law was able, for the first time in the seventy-two-year history of the Soviet form of socialism, to end one-party rule and win control of the government. In the Soviet Union, the democratic impulse took a more subtle form, creating in the minds of the generation exemplified by Mikhail Gorbachev the vision and the will to overcome, from within, the oppressive antidemocratic regime under which they had matured. In China this same impulse, expressed only as an inchoate demand for something called "democracy"—more felt than understood—led to the tragedy in Tiananmen Square, but nevertheless demonstrated to the world the impelling force of the idea. In the United States and Europe, perhaps mollified to a degree by the absence of formal constraints on public discourse, the demands voiced by the autonomous popular movements have been more narrowly defined and less dramatic, yet effective. They have forced partial disarmament on the great powers, and everywhere at least the recognition of the environmental crisis. In the United States, these movements have won major governmental concessions: civil rights, women's rights, the right of sexual preference.

The universality of its expressions—in the march on Selma, Alabama; the students occupying Tiananmen Square; the crowds pouring into the Potsdamer Platz and Wenceslas Square; the huge demonstrations for nuclear disarmament in Aldermaston, England, or Washington, D.C.; the strike at the Lenin Shipyards in Gdansk; or for that matter, decisions made by a few men in the Kremlin—confirms the existence of a common desire for everyday democracy that is deeply felt, powerful, and eventually undeniable. Moreover, it is an impulse that generates novelty: inventions as simple as leaving one's country, as unplanned as American mothers' boycott of apple juice, as subtle and designed as *glasnost* and *perestroika.*

Recent history is a powerful assurance that the common

democratic impulse can be translated into new, practical forms. Living in such a moment in history, it would be unwise—and indeed unworthy—of us to dismiss as an illusion the idea that environmental and economic democracy can be realized by implementing the social governance of production.

It is useful to remember that there are generals and reluctant foot soldiers in the technosphere's war against the ecosphere. Pogo's analysis of the environmental crisis—"We have met the enemy and he is us"—is appealing but untrue. It is true, of course, that the householder who discards nearly two hundred pounds of plastic trash annually has struck a blow against the ecosphere. But often the householder has no choice; after all, milk is no longer sold in returnable glass bottles. Moreover, the decision to produce the plastic in the first place was not made by the householder, nor does the householder benefit from the profits that motivated that decision. That power and that motivation reside with the corporate managers—the generals who order the assault, unwittingly, on the ecosphere. This narrow segment of society makes the decisions that obligate the rest of us to participate in the ecological war. And it is the corporate generals who reap the short-term—and shortsighted—economic benefits. It is, of course, always possible that, appalled by the horrors of ecological warfare, the generals may overcome their greed for greater profits and call off the attack. But the record suggests that this is very unlikely. For twenty years or more, like the rest of us, the corporate managers have known about the catastrophic ecological consequences of their decisions. But they have usually responded by bitterly resisting almost every effort to even acknowledge, let alone reduce, the environmental impact generated by those decisions. Given that record, it is unlikely that they will voluntarily relinquish their powerful positions and call off the war against nature.

If, as I suggest, the corporate managers represent a small

but exceedingly powerful minority whose private interests conflict with the majority's interest in ecological peace, where can we find the means to diminish their power? The proper answer, I believe, is the traditional one: in the power of a democratic government. The effort to achieve political power in order to use it to solve the environmental crisis—is well under way in Europe, and less dramatically in the United States as well.

The most striking example is given by the brief but remarkable history of the Green Party in West Germany. In 1983, only four years after it was founded, the Green Party obtained 5.6 percent of the vote in the national election and won seats in the Bundestag. The party began with a purely ecological outlook, which emphasized not only the hazards of pollution, but also the dangers of the ultimate ecological disaster—nuclear war. The Greens' antiwar position was an important source of votes, for neither the then reigning Social Democrats nor the opposition Christian Democrats were opposed to the expanding U.S. nuclear presence in West Germany and Europe. In the 1983 election, peace-oriented voters, who were not necessarily preoccupied with ecological issues, found in the Greens a way to express this concern. This helped the Greens appeal to a broader segment of the voting population than just ecological activists.

On the other hand, the Greens' preoccupation with ecology was also a serious political drawback. The 1983 election occurred at a time of high unemployment, unprecedented in recent German history. Some Greens believed that the best way to reduce environmental degradation is to lower the level of industrial activity, which they regarded as inherently antiecological. In turn, this belief was translated into an open disregard for the demand for more jobs; after all, more jobs means putting more people to work in polluting factories. Naturally, unemployed workers as well as many other voters who sympathized with them were not attracted to the Greens.

Thus, from its inception, like the environmental movement generally, the Green Party ran head-on into the firm bond between ecology and economics. This confrontation gave rise to a basic split within the Green Party. One group, the "fundamentalists," adhere to what they regard as the basic precepts of the ecological orientation: a social structure based on "unity with nature"; a spiritual devotion to harmonious relations among people and nations; respect for all forms of life. (At a Green convention, one member of this group pleaded passionately for a resolution condemning the "slaughter" of thousands of frogs in biology laboratories, comparing it to the Holocaust.) The fundamentalists are chiefly concerned with combatting pollution, fostering a transition to renewable resources, alternative "soft" technologies, and a life-style appropriate to them.

The other group is the "realists." They recognize the fundamental origin of ecological and other problems in the political control of the economy. They are concerned not only with the technology of production and its impact on both the environment and jobs, but also with deciding which social class holds the governing power. The realists are linked to labor unions and most of them favor political alliances, for example with the Social Democrats, whose positions on environmental and peace issues have in some respects now moved closer to the Greens' position.

In October 1985, the split between the fundamentalists and the realists took a dramatic turn, when, over the objections of the national Green leadership, the realist-led Green Party of the industrial state of Hesse agreed to join the Social Democrats in a governing coalition. The Christian Democrats promptly attacked the Hesse Social Democrats for making a pact "with fanatical opponents of our free economy." In the heat of the accompanying debate, the fundamental political issue generated by ecological concerns became plain when a Christian Democrat who opposed the coalition declared that major corporations would leave

Hesse because "free companies decide where they want to invest." In the aftermath of the Chernobyl disaster, the Social Democrats have moved much closer to the Green position against nuclear power, and polls suggest that an alliance of the two parties might win the next federal election.

In the last few years, the "realist" approach has largely governed Green Party politics and regional alliances with the Social Democrats have expanded. Particularly important is the Green–Social Democratic coalition that took power in West Berlin in 1989. Thus, the Green Party in West Germany has become a lively and effective dramatization of the links that tie ecology to economics and both to politics. It also illustrates the ease with which a failure properly to perceive these links can lead to a politics which rises in defense of laboratory frogs but not unemployed workers.

Italy provides a different example of environmental politics. While the Italian Greens are strongly oriented toward environmental issues—in particular opposition to nuclear power, nuclear weapons, and industrial pollution—they share the general political orientation of the Italian parties of the left. This leads to a kind of "Red-Green" phenomenon in which environmental concerns are joined to a broader left political program. In this sense, the Greens and environmentalists affiliated with the left parties also become important through the pressure they exert on the parties to adopt environmental positions—for example, a few years ago helping to reverse the Italian Communist Party's support for nuclear power. In turn, such developments encourage environmentalists to work within the parties, thus linking environmental concerns more firmly to the rest of politics. In the 1988 parliamentary elections, running on either one of two Green slates or on the Communist Party and Socialist Party slates, some 40 leading environmentalists were elected to the 630-member parliament. Even larger votes were cast for Green candidates in the 1989 elections to the European Parliament.

In the United States, environmental politics has followed a distinctive course. Here the environmental movement is part of the phenomenon which, since World War II, has given rise to wave after wave of popular, issue-oriented movements: for civil rights; against nuclear weapons testing; for women's, gay, and lesbian rights; against the war in Vietnam; for the environment; against nuclear power and for solar energy; for world peace. These movements have a good deal in common. All of them have arisen outside the arena of conventional politics, sparked by outsiders like Martin Luther King, Jr., and Rachel Carson, rather than by established political figures. Their level of public support has typically gone through successive cycles of enthusiasm and apathy. At their height the movements have achieved notable successes—the civil rights and environmental laws, the nuclear test ban treaty, the new employment opportunities for women—all of them accomplished by nonelectoral means: marches, demonstrations, and massive lobbies. Yet, as the record of the Reagan administration shows, these accomplishments can be quickly eroded when officials who are hostile to them are elected to power. Indeed, the movements' greatest failure has been their inability to translate the millions of votes that their combined adherents represent into significant electoral power and thereby elect people to office who will protect their gains and expand them.

In recent years, American political life has not been hospitable to such issue-oriented confrontations. Electoral campaigns have become hugely expensive advertising extravaganzas rather than intense, ongoing discussions of the issues. (When I ran for president in 1980 as the Citizens Party candidate, I was asked by a television reporter: "Are you a *serious* candidate, or are you just running on the issues?") Such discussions as have occurred are themselves largely segregated from the campaigns and confined to the hollow formality of "debates" in which journalists sometimes play a larger role than the candidates. Even the Congress—

which has a constitutional duty to debate the relative merits of government expenditures—has abdicated that responsibility by passing the Gramm-Rudman law. Like an environmental risk-benefit analysis, the Gramm-Rudman law substitutes mathematical formulas for moral judgments; it legitimizes the growing trivialization of politics.

A major reason for the tenuous connection between the movements and politics is that their issues have been consigned to the political ghetto that is reserved for "special interests." In a sense, this isolation is self-imposed, for the varied concerns are usually manifested—for example in a legislator's office—as single-minded constituents, one pleading for peace, another for sexual equality, a third for environmentalism, on down the list of social issues. Each special pleading demands a special response, at best unrelated to the other issues, but often in conflict with them. Each of the issues is regarded as a possible modifier, but not a creator, of national policy—a course correction in the trajectory of the ship of state, but not a motive force. Yet, taken together, and added to the much older labor movement, the issues that the movements represent comprise not only the major aspects of public policy, but its most profound expressions: human rights, the quality of life, health, jobs, peace, survival. What could bring these movements together, enabling them to exert an effect on national policy that expresses the deep political meaning of their collective concerns?

The environmental experience suggests an answer. As we have seen, the obvious manifestations of the environmental problem—smog, toxic dumps, or nuclear catastrophes—which set it apart as a "special interest" are but the perceptible expressions of a deeper, underlying issue: how the national system of production is to be governed. Here environmentalism reaches a common ground with all the other movements, for each of them also bears a fundamental relation to the governance of production. For example, like pollution, a major feature of discrimination—paying women and racial minorities less than white men—originates in decisions made by the managers of productive enterprises. Of course, other social, cultural, and psychological factors are involved as well, but the end result—wage discrimination—is, after all, an effective way of reducing production costs. Or, to look at this relationship the other way around, if it were determined through some system of social governance that all the employees in an enterprise who do comparable work should receive equal pay regardless of sex or race, a major effect of the social, cultural, and psychological forces that engender discrimination would be largely nullified. The connection between this common ground and the issues of peace and foreign policy is less direct, but nevertheless substantial. For example, it explains the apparent justification for the U.S. government's military belligerence in recent years—that force must be used wherever it is needed to support governments and political groups that, like the United States, favor "free-enterprise" principles of economic governance.

In sum, perhaps the most useful outcome of the environmental experience is that it illuminates the relationship between the outward manifestations of the ills that trouble modern society and their common origin. But there are risks in expounding this relationship. Calling attention to their source may appear to minimize the importance of the immediate problems that initially attract adherents to the cause, risking their ire. At the same time, an effort to transform the "special interest" into a critique of basic, even more troubling faults in the social structure is likely to generate more intense opposition. The path taken by Martin Luther King, Jr., in the last few years of his life is a cautionary example. At the height of his influence, King had won major victories and had acquired a broad following as the leader of a powerful attack on the outward expressions of legally enforced racial discrimination: segregated schools and public facilities. Then, a few years before he was assassinated, King began to link racial discrimination to its origins and, thereby, to other social issues. He sensed, it seems, that blacks could not break out of their persistent social ghetto if they remain trapped, as a "special interest," in the economic ghetto of poverty. And so he led a march of poor people (black and white) on Washington, championed the cause of striking garbagemen in Memphis, and declared his opposition to the war in Vietnam. In this new role, King quickly became more controversial. He was out of his depth, it was said, diverted from his true mission and taken in by political radicals. But he died believing, it would seem, that this new course had brought him closer to the heart of the problem that he had set out to resolve; that beneath the legal basis of racial discrimination lay the deeper problems of poverty and violence; that the root of racial discrimination is also the root of poverty and war. The Rainbow Coalition, which evolved out of Jesse Jackson's entry into politics, can be seen as a current effort to continue on this road.

Neither the environmental movement, nor any of the other issue-oriented movements, has yet found the path that Martin Luther King began to travel before he died. But the environmental experience powerfully illuminates the direction that it must take.

What the United States—or indeed any one country—does will not in itself end the war against nature. It is a global war and only global action will end it. Even concerted action by the northern industrialized countries—where most of the assault on the ecosphere now originates—will not be enough. What is done in developing countries is crucial as well. As we have seen, environmental improvement is intimately linked to economic development. As the world population doubles in the next forty years or so—90 percent of the increase in developing countries—worldwide production levels will need to increase sharply in order to sustain economic development in the Third World. Unless the expanded production facilities are ecologically sound, this process will further degrade the environment.

There are serious constraints on developing countries that, if unrelieved, will greatly reduce their ability to participate in the transition to ecologically sound systems of production. Since for some time the required production facilities—for example, solar energy equipment—would need to be imported, foreign credit will be badly needed. Yet the developing countries' current debt to northern banks—about $1 trillion—is so large that the interest payments alone condemn most of them to an economic treadmill. At present, some $30 billion a year flows from the poor, developing countries to the rich developed ones. The only conceivable way of removing this irrational barrier to the ecological transition is to eliminate these debts. I believe that this should be regarded not as a magnanimous gesture but as partial reparations for the damage inflicted on the developing countries by the former colonial empires.

Finally, there is the calamitous fact that many developing countries have been wracked by wars which, apart from their destructive impact, have consumed a significant part of their foreign credits in arms purchases. Most of the people in the Third World have been desperately poor and ruled by a small, relatively rich minority, native or foreign. It is an imperative of human history that poor, disenfranchised people do not indefinitely endure this condition. Inevitably, they seek, by force if need be, to govern the nation's resources and its political power more equitably. Each embattled nation then becomes an arena for competition between the United States and the Soviet Union, which provide not only the arms, and sometimes the troops, but also the ideological banners under which the wars are fought. All over the world, the United States has sold weapons on the promise that they would be used to oppose Soviet influence, and Korea, Vietnam, and Cambodia have felt the deadly force of the U.S. war machine itself. The Soviet Union has responded similarly—in Afghanistan, Ethiopia, and Angola—to bolster their side. The result is the terrifying fact that since the end of World War II, some 25 million people have died in wars—almost without exception in the Third World. The wars have been fought almost entirely in the lands of people who are black, Asian, or Hispanic. In a sense, the United States and the Soviet Union have been fighting World War III by proxy—using Third World peoples like pawns in a war game.

The two countries have seen each other as enemies since the revolution that created the Soviet Union in 1917, albeit with varying levels of intensity and expression. Following the temporary truce in World War II, this enmity has largely conditioned the state of the world. It has enveloped the world in the threat of nuclear annihilation; it has saddled both countries and their allies with the economic drain of monstrous military budgets; it has at least facilitated and often generated devastating wars in the Third World. As a result, about $1 trillion is spent worldwide annually to prepare for war or to engage in it, diverting this huge sum from the urgent task of relieving poverty, sickness, illiteracy, and environmental degradation.

As we have seen, in the United States the reconstruction of the major systems of production that will be required to resolve the environmental crisis will cost about $100 billion per year for at least a ten-year period. In order to approximate the cost of this transformation worldwide, we might expand the U.S. figure in line with the relative sizes of the American and world GNPs. Since the United States accounts for about 26.8 percent of the world GNP, the worldwide cost of the necessary transformation of the major production systems would be about $370 billion per year for ten or more years. As already indicated, the global environmental crisis cannot be resolved without the participation of Third World countries, which in the process must increase their present rate of economic development. At the least, this will require retiring the $1 trillion Third World debt over the crucial ten-year period—an added annual cost of about $100 billion. This brings the total cost of the global transition to an ecologically sound system of production to perhaps $500 billion per year for ten years or more.

Obviously, given the uncertainties about the actual costs of the technological transformation in the United States and their extension to a worldwide figure based, crudely, on relative GNPs, this cost is very approximate. A similar exercise by the Worldwatch Institute, which assumed only a 30 percent reduction in the Third World debt, estimates the total world cost of the transition at an average of $137 billion annually over a ten-year period.

Although these numbers are only rough approximations, they can help to place the cost of the environmental transition in a realistic economic framework. Clearly, the sums are so large that some existing expenditures will need to be reduced in order to accommodate the environmental transition. The obvious candidate is the approximately $1 trillion now spent annually, worldwide, for military expenditures. There are several reasons for this conclusion. The first reason is simply numerical: the military budget is large enough to serve the purpose. The second reason is in the realm of social justice, or morality: it would be only right to divert these sums from a destructive—or, at the least, a grossly unproductive—purpose to the creation of new ecologically sound and economically beneficial systems of production. The final reason is timeliness: with nuclear disarmament already well under way and the sudden end of East-West tensions in Europe, military needs, by any reasonable measure, have been sharply reduced. Already the Soviet Union has announced an approximately 14 percent reduction in its military budget, and in the United States, for the first time since World War II, the Pentagon has proposed ways of reducing military expenditures by some $180 billion over the next few years.

Thus, resolving the environmental crisis would require a let us say 50 percent cut in worldwide military expenditures. In sum, we can end the environmental crisis by sharply reducing the present global commitment to militarism. This is not only a matter of money, for military expenditures divert resources from constructive uses in particularly crucial sectors of the economy. One of these is research and development, services that are vital to the transformation of production systems. Although military expenditures represent only about 6 percent of total world production, military research and development has been growing at twice the rate of overall military expenditures. As a result, the military has now commandeered the services of about one-half of the world supply of scientists and engineers.

Military expenditures also degrade a nation's productive capacity, reducing its ability to develop economically. This is suggested by empirical evidence about the relationship between military expenditures and a major index of economic strength—the annual increase in economic productivity. Among the industrialized nations, there is an inverse relationship between the rate of productivity increase and the country's military budget, expressed as a percentage of its GNP. Thus, over the last decade, the highest average rate of increase in economic productivity, about 3 percent annually, has occurred in Japan, where, in keeping with the terms of the peace treaty, the military budget has been held to 1 percent or less of the GNP. At the lower end of the productivity scale, with an average annual increase of only 0.5 percent, we find the United States, where the military budget has been about 7 percent of the GNP. Unburdened by a huge military budget, over the last decade Japan has been able to use an average of about 30 percent of its GNP as business investment capital, whereas in the United States this figure is only about 16 percent. Although statistical data are lacking, it is evident that the Soviet economy has also suffered from the impact of its very large military budget.

Thus, military costs hinder economic productivity by diverting capital from productive investments. In effect, military expenditures tend to damage the economy at its points of *growth*—that is, investment in new productive facilities. This is understandable. If, for example, an airline company purchases a jumbo jet, the plane becomes an instrument of production, generating wealth as it flies. On the other hand, if the U.S. Air Force purchases that same airplane, it will produce nothing of economic value.

There is, therefore, a strong link between ecological reconstruction and disarmament. One of the important corollaries of disarmament is the conversion of military facilities—and their technical and scientific personnel—to peaceful purposes. This could make a major contribution to the transformation of the present environmentally hazardous systems of production into ecologically sound ones. And it is only fitting that ecological reconstruction, a process that enhances both environmental quality and economic productivity, should be supported by funds diverted from the military, which, in its reliance on nuclear war, is an enormous environmental threat and which, in its economic impact, is so destructive.

In sum, war and the threat of war are major obstacles to resolving the environmental crisis. If the militarism that has gripped the world since World War II continues, it will effectively block the huge and costly effort that must be made to end the environmental crisis. The major force that has generated global militarism is the long-standing enmity between the United States and the Soviet Union. Although the most virulent aspect of this hostility—the threat of nuclear war—has greatly diminished in the last few years, the underlying antagonism persists and continues to condone the huge military expenditures that block the hoped-for ecological transformation. What, then, stands in the way of at last ending this costly enmity between the United States and the Soviet Union?

The environmental crisis seems to illuminate even this somber issue. Central to the effort to resolve the crisis, in both the United States and the Soviet Union, is their approach to the crucial question of social versus private governance of the systems of production. To the U.S. government—whether openly expressed or inferred—the Soviet Union has been, in Ronald Reagan's inflammatory words, the "evil empire," bent on destroying the U.S. bastion of capitalism. To the Soviet government, the United States has been the powerful enemy that has made persistent political and military efforts to contain the spread of socialism from its beginnings in the Bolshevik revolution. Until Gorbachev's rise to power, this was a relentless conflict, which was regarded as the motive for the nuclear war that for forty-five years each side thought the other would unleash. Now in only a few years these rigid postures have changed, marked, in real terms, by dramatic arms reduction agreements and by Gorbachev's apparent intention to avoid the use of force, even to control his own socialist allies.

On the ideological front, the rapid dynamics of *glasnost* and *perestroika* have openly raised the issue of the governance of production in the Soviet Union. In September 1989, the new Soviet legislature began an extensive debate on the question of private property, specifically raising the question of whether it should include not only small family enterprises but major industrial operations as well. Thus far, the answer appears to be no. More important is that the role of social governance of production—which is crucial to solving the environmental crisis—is now openly discussed.

The United States is still in the grip of the rigid taboo against even questioning—let alone changing—our commitment to exclusively private governance of production. Yet, under the pressure of our monumental failure, after a twenty-year effort, to improve the environment, there are a few embryonic signs of motion. If these small movements grow, and we in the United States begin our own *perestroika,* albeit starting from an opposite ideological position, it may be possible to appreciate that we occupy a common ground with the Soviet Union. It may become clear to both of us that the ideological grounds for our mutual hostility, if not yet destroyed, have become too blurred and ambiguous to justify the enmity that has threatened and impoverished the world.

We can now see that both our suicidal assault on nature and the wars and the threat of wars that have engulfed the world in misery have a common origin: the failure, in capitalist and socialist countries alike, to begin a new historic passage—toward a democracy that encompasses not only personal and political freedom, but the germinal decisions that determine how we and the planet will live. Now that the people of the world have begun to understand that survival depends equally on ending the war with nature and on ending the wars among ourselves, the path to peace on both fronts becomes clear. To make peace with the planet, we must make peace among the peoples who live in it.

NOTES

CHAPTER 1. AT WAR WITH THE PLANET

page 5: temperature data from analysis of Antarctic ice: See article on global climatic change by R. A. Haughton and G. M. Woodwell in *Scientific American,* April 1989. Numerous articles and reports on this subject have appeared recently; this one is particularly comprehensive.

page 6: historic changes in carbon dioxide levels and temperature: These data are summarized in the article cited above.

page 8: the basic laws of the ecosphere: The informal "laws" that describe the behavior of ecosystems are elucidated in some detail in my earlier book *The Closing Circle* (New York: Knopf, 1971).

page 9: engine design and smog: Chap. 4 in *The Closing Circle* (cited above) is a detailed account of this issue.

page 9: nitrate pollution from fertilizer: Chap. 5 in *The Closing Circle* is a detailed account of this issue.

page 11: Walter Elsasser's computation: See his interesting work *Atom and Organism: A New Approach to Theoretical Biology* (Princeton, N.J.: Princeton University Press, 1966).

page 12: very few chlorinated organic compounds: See article by D. John Faulkner in *Tetrahedron Report* no. 28 (Elmsford, N.Y.: Pergamon Press, 1977).

page 15: President Bush's scheme: This is contained in his new Clean Air Bill, as announced in June 1989.

page 17: Earth First! article on AIDS: See *Utne Reader,* Dec. 1987.

CHAPTER 2. THE ENVIRONMENTAL FAILURE

page 21: considerable cut in CEQ budget: See *New York Times,* Sept. 29, 1981.

pages 21–2: air pollution trends: Data on emissions and local concentrations of standard air pollutants are from *National Air Quality and Emission Trends Report, 1984,* U.S. EPA Office of Air Quality and Standards, Research Triangle Park, N.C., April 1986 (for 1975–1984 data), and *National Air Quality and Emission Trends Report, 1987,* U.S. EPA Office of Air Quality and Standards, Research Triangle Park, N.C., March 1989 (for 1984–1987 data).

page 22: lead levels in children's blood: See article by S. L. Annest et al. in *New England Journal of Medicine,* June 8, 1983.

page 23: air quality in Los Angeles and other cities: See *Environmental Quality,* 15th Annual Report of the Council on Environmental Quality, Washington, D.C., 1984.

pages 23–4: effects of acid rain: For a general description of the effects, see National Research Council, *Acid Deposition: Atmospheric Processes in Eastern North America* (Washington, D.C.: National Academy Press, 1983). The apparent effects of acid rain on the increased mortality of trees and their reduced growth rate in the Southeast was reported by Philip Shabecoff in *New York Times,* Dec. 12, 1985.

page 24: acidity data from Hubbard Brook: See National Research Council, *Acid Deposition.*

page 25: U.S. Geological Survey report on water pollution trends: See R. A. Smith et al. in *Science,* vol. 235, p. 4796.

page 25: bacterial count too high to permit swimming: The data are in the USGS report cited above. For the relevant standards, see E. T. Chanlett, *Environmental Protection*, 2nd ed. (New York: McGraw-Hill, 1979).

page 27: eutrophication of Lake Erie: Chap. 6 of Barry Commoner, *The Closing Circle* (New York: Knopf, 1971), discusses the state of the lake in 1970.

page 27: trends in Lake Erie shoreline eutrophication, oxygen depletion, phosphate, and fish catch: See *Lake Erie Intensive Study, 1978–1979: Final Report,* Great Lakes National Program Office, U.S. EPA Region 5, Chicago, 1984.

page 28: 1982 survey of lakes: See National Water Quality Inventory, 1982 Report to Congress, U.S. EPA, Washington, D.C., 1984.

page 28: quality of groundwater: The results of the USGS well tests, including observations in Nebraska and the Sacramento Valley, are presented in *The National Water Summary, 1984,* U.S. Geological Survey Water Supply Paper no. 2275, Washington, D.C., 1985.

page 29: USGS quotation: See USGS report cited above.

pages 29–30: reduced levels of DDT and PCB: See *Environmental Quality, 1987–1988,* 18th and 19th Annual Report of the Council on Environmental Quality, Washington, D.C., 1989, pp. 417, 319–26.

page 30: synthetic chemicals in body fat and breast milk: The body-fat data are reported in *Broad Scan Analysis of the FY82 National Human Adipose Tissue Survey Specimens,* U.S. EPA Office of Toxic Substances, Washington, D.C., 1986. Data on the occurrence of various toxic synthetic organic chemicals in breast milk are presented in *Acquisition and Chemical Analysis of Mother's Milk for Selected Toxic Substances,* U.S. EPA Office of Toxic Substances, Washington, D.C., 1980. Data on the pesticides present in breast milk in the United States are reported in *Na-*

tional Study to Determine Levels of Chlorinated Hydrocarbon Insecticides in Human Milk, U.S. EPA, Washington, D.C., 1976 (supplement in 1977). This report also reviews similar studies in Europe, Australia, Japan, and New Guinea. Levels up to thousands of parts per billion are reported wherever chlorinated pesticides are in use. A. Schecter et al. have reported on dioxins and furans in breast milk in the North American continent and Vietnam in *Chemosphere,* vol. 16 (1987), p. 2003. A paper presented by A. Schecter et al. at the Dioxin '89 Conference in Toronto, Canada, reports on dioxin in human milk from the Soviet Union. Another paper from the same author presented at Dioxin '89 reports on dioxin in human milk from Africa, Pakistan, and England.

page 30: dioxin's extraordinary ability to enhance incidence of cancer: The cancer risk resulting from exposure to dioxin was exhaustively analyzed in *Health Assessment Document for Polychlorinated Dibenzo-p-Dioxins,* U.S. EPA, Washington, D.C., 1985.

page 30: dioxin in adipose tissue: See *Broad Scan Analysis of the FY82 Human Adipose Tissue Survey Specimens.*

page 30: greater than maximum lifetime cancer risk from benzene: The maximum lifetime cancer risk to the U.S. population due to exposure to benzene is 154 per million, according to W. Hunt et al., *Estimated Cancer Incidence Rates for Selected Toxic Air Pollutants Using Ambient Air Pollution Data,* U.S. EPA, Washington, D.C., 1985.

page 31: origin of dioxin in trash-burning incinerators: See chapter 6 below for a discussion of this issue.

page 31: dioxin analysis of dated Great Lakes sediments: See J. M. Czuczwa and R. A. Hites in *Environmental Science and Technology,* vol. 18 (June 1984), p. 844.

pages 31–2: estimates of toxic chemical emissions: The Toxic Release Inventory (TRI) was carried out in 1988 by EPA under authority of Sec. 313 of the Emergency Planning and Right-to-Know Act. It was designed to provide the public with information about toxic substances released by industrial establishments with more than 10 workers. The survey reported that about 2.2 billion pounds of toxic chemicals were released annually, representing some 300 different substances. In testimony before the Subcommittee on Superfund, Ocean and Water Pollution, Committee on Environment and Public Works, U.S. Senate, on May 10, 1989, Dr. Joel Hirschhorn of Congress's Office of Technology Assessment evaluated these results. His testimony points out that the emissions were underreported and that many facilities with fewer than 10 workers handle toxic chemicals. On the basis of such inadequacies, he suggests that the TRI data are 20 times too low.

page 32: only 1 percent of toxic waste destroyed: See *National Survey of Hazardous Waste Generators and Treatment, Storage and Disposal Facilities Regulated Under RCRA in 1981,* U.S. EPA, Washington, D.C., 1984.

page 32: increased birth defects in Great Lakes birds: For accounts of these studies, see *New York Times,* July 12, 1988, and March 21, 1989.

page 32: toxic compounds in adipose tissue: See *Broad Scan Analysis of the FY82 National Human Adipose Tissue Survey Specimens,* vols. 2 and 3.

page 32: National Research Council report: See National Research Council, *Regulating Pesticides in Food: The Delaney Paradox* (Washington, D.C.: National Academy Press, 1987), p. 49.

pages 32–3: hazards from nuclear weapons production: For many years the U.S. government flatly denied there were any hazards from the production of nuclear weapons. Only in the last few years has the government been forced to acknowledge what was already documented in secret reports—that nearly all of the production facilities had released dangerous amounts of radiation. One facility, at Rocky Flats, Colo., is so badly contaminated that it must be closed. For an excellent account of these problems, see John May, *The Greenpeace Book of the Nuclear Age* (New York: Pantheon, 1990), pp. 301–8. This book is a valuable source of comprehensive information about nuclear hazards.

page 33: radiation exposure from normal nuclear plant operations: See *The World Environment, 1972–1982,* United Nations Environment Programme, New York, 1982, pp. 471–75. Apart from accidents, about 78% of the average human exposure to radiation is from natural sources; about 21% is from medical sources; and about 0.1% derives from the normal operation of the nuclear power industry. However, such average data may well include significantly higher exposures of those people living near nuclear power plants. The possibility that "normal" emissions of radioactivity from nuclear power plants may be responsible for increased health hazards, especially to infants, is considered in J. M. Gould and B. A. Goldman, *Deadly Deceit* (New York: Four Walls Eight Windows, 1990).

pages 33–4: krypton 85 from nuclear power plants: See *Environmental Quality, 1987–1988,* p. 419.

page 34: Chernobyl nuclear power plant disaster: See C. Hohenemser, "The Accident at Chernobyl," *Annual Review of Energy,* vol. 13 (1988), p. 383, and C. Flavin, *Reassessing Nuclear Power: The Fallout from Chernobyl,* Worldwatch Paper no. 75, Worldwatch Institute, Washington, D.C., 1987. Also see the account in May, *Greenpeace Book of the Nuclear Age,* pp. 280–92.

page 36: reduction in strontium 90 levels: See *Environmental Quality, 1987–1988,* p. 418.

page 36: European environmental data: For the German data see H. Weidner, *Air Pollution Control Strategies and Policies in the F.R. Germany* (Berlin: edition sigma, 1986), p. 19. For the British data see H. Weidner, *Clean Air Policy in Great Britain* (Berlin: edition sigma, 1987), p. 37.

page 36: air quality in Poland: See *Environment Statistics in Europe and North America,* United Nations and Economic Commission for Europe, New York, 1987, p. I-30.

page 36: recent report from Global Environment Monitoring System: The report is summarized in *Environment,* vol. 31, no. 8 (Oct. 1989), p. 6.

page 37: pollution in Baltic Sea: See *Environment Statistics in Europe and North America,* p. II-64.

page 37: steps to reduce CFC production: The Montreal Protocol, signed by 24 countries in September 1987, calls for a reduction in CFC consumption by 20% in 1993 and 50% by 1998, as compared with the 1986 level. However, even a 50% reduction will result in a quadrupling of man-made chlorine concentrations. Much more stringent actions are needed. These are described in A. Makhijani et al., *Saving Our Skins,* Environmental Policy Institute and Institute for Energy and Environmental Research, Washington, D.C., Sept. 1988.

page 38: most recent CEQ report: See *Environmental Quality, 1987–1988,* pp. 328–29, 337.

page 38: forest clearing in Latin America: See L. R. Brown et al., *State of the World, 1988* (New York: Norton, 1988), p. 103.

pages 38–9: rule-making document on air quality standards: The document, "National Air Quality Standards," is published in *Federal Register,* vol. 36, pt. 2 (April 7, 1971), pp. 6680–6701.

page 39: nearly 150 million people still breathe substandard air: See *National Air Quality and Emission Trends Report, 1987,* U.S. EPA Office of Air Quality and Standards, Research Triangle Park, N.C., March 1989, p. 5.

CHAPTER 3. PREVENTION VERSUS CONTROL

Page 43: efficiency of controls: For the efficiency of sulfur dioxide controls, see T. C. Elliott et al., *The Acid Rain Sourcebook* (New York: McGraw-Hill, 1984), p. 171. For the efficiency of carbon monoxide controls, see F. P. Grad et al., *The Automobile and Regulation of Its Impact on the Environment,* (Norman: University of Oklahoma Press, 1975), p. 121.

page 43: tests of catalytic converters: See *Environmental Quality,* 15th Annual Report of the Council on Environmental Quality, Washington, D.C., 1984, p. 71.

page 44: mercury in Lake Erie sediments: See *Lake Erie Intensive Study, 1978–1979: Final Report,* Great Lakes National Program Office, U.S. EPA Region 5, Chicago, p. F-81.

page 45: toxaphene content of fish: See *Environmental Quality,* p. 599.

pages 46–7: post-1950 changes in production technology: The detailed data are presented in Barry Commoner, *Chemistry in Britain,* vol. 8, no. 2 (Feb. 1972).

page 48: automobiles and trucks responsible for 80 percent of lead emissions: See *National Air Pollutant Emission Estimates, 1940–1986,* U.S. EPA Office of Air Quality and Standards, Research Triangle Park, N.C., 1988, p. 44.

page 49: changes in farm output and use of labor, machinery, nitrogen fertilizer, and pesticides, 1950–1970: These data are from *Economic Report of the President* (Washington, D.C.: Government Printing Office, Jan. 1989). The output data are from table B-97, p. 419; the input data are from table B-98, p. 420.

page 50: seven hundred synthetic chemicals introduced into agriculture: See National Research Council, *Regulating Pesticides in Food: The Delaney Paradox* (Washington, D.C.: National Academy Press, 1987), p. 52.

page 51: National Research Council assessment of cancer from pesticide residues in food: See NRC report cited above, p. 57.

page 51: National Cancer Institute study of cancer in Kansas farmers: See D. P. Buesching et al., "Cancer Mortality Among Farmers," *Journal of the National Cancer Institute,* vol. 72 (1984), p. 503.

page 51: annual output of chemical industry 500 to 600 billion pounds: See *Chemical and Engineering News,* June 19, 1989.

page 52: chemical disaster in Bhopal, India: For an enlightening account of this disaster, see L. Everest, *Bhopal: The Anatomy of a Massacre* (New York: Banner Press, 1986).

page 52: repeated accidents at Union Carbide methyl isocyanate plants: According to a report issued by U.S. EPA, the Union Carbide plant in Institute, West Virginia, experienced 28 accidental leaks of methyl isocyanate in a five-year period. See *New York Times,* Jan. 24, 1985.

page 53: parathion applied to El Salvador coffee crop: See *Better Regulation of Pesticide Exports and Pesticide Residues in Imported Food Is Essential,* Report to Congress, U.S. General Accounting Office, Washington, D.C., 1979, p. 8.

page 53: India's use of pesticides banned in U.S.: See GAO report cited above, p. 89.

CHAPTER 4. THE COST OF FAILURE

page 57: cost of environmental control program: $365 billion in last five years: See *Survey of Current Business,* U.S. Department of Commerce, June 1989, pp. 24–25.

page 58: air pollution rule-making document: This is published as "National Air Quality Standards" in *Federal Register,* vol. 36, pt. 2 (April 7, 1971), pp. 6680–6701.

page 60: National Environmental Policy Act purpose, quotation: The quotation is from Sec. 2 (Purpose) of the National Environmental Policy Act of 1969.

page 60: CBS public opinion poll, quotation: See *New York Times,* July 2, 1989.

page 61: Bhopal disaster, quotation: The quotation is from the introduction to *Time* magazine's cover story on the Bhopal disaster, Dec. 17, 1984, p. 20.

page 62: EPA estimate of dioxin cancer risk: See *Health Assessment Document for Polychlorinated Dibenzo-p-Dioxins,* U.S. EPA Office of Health and Environmental Assessment, Washington, D.C., 1985.

pages 62–3: American Industrial Health Council on comparative risks of death, quotation: See *Chemical and Engineering News,* Jan. 30, 1978, p. 30.

page 64: National Peach Council on DBCP risk/benefit, quotation: See *New York Times,* Sept. 27, 1977.

page 64: OMB definition of risk management, quotation: See *Regulatory Program of the United States Government, April 1, 1986–March 31, 1987,* Executive Office of the President, Office of Management and Budget, 1987, p. xxi.

page 65: OMB on cost-efficient regulation, quotation: See OMB report cited above, p. xxi.

page 65: cost of shutting down cigarette industry: The value of the shipments of cigarettes by the U.S. tobacco industry is given as $17.3 billion (see *U.S. Industrial Outlook, 1989,* U.S. Department of Commerce). That income provides the return on investment and payments to industry workers and tobacco growers.

page 66: Reagan's executive order: See *Regulatory Program of the U.S. Government,* p. xix.

page 66: most toxic dumps located near black and Hispanic communities: See *Toxic Wastes and Race in the United States,* Commission for Racial Justice, United Church of Christ, New York, 1987.

page 67: EPA cost-benefit analysis of removing lead from gasoline: See *Environmental Quality,* 15th Annual Report of the Council on Environmental Quality, Washington, D.C., 1984, pp. 77, 235.

page 68: New York State Department of Environmental Conservation's remarkable pronouncement: See *New York Times,* Sept. 30, 1987.

page 69: OMB on risk of chemically induced cancer, quotations: See *Regulatory Program of the U.S. Government,* p. xxiii.

page 69: "A more accurate estimate . . .", quotation: See *Regulatory Program of the U.S. Government,* p. xxiv.

page 70: AAF must be acted on by liver enzyme: See E. C. Miller and J. A. Miller in *Chemical Carcinogens,* American Chemical Society Monograph no. 173, 1976.

page 72: EPA scientific committee echoing OMB's call: For an expression of the view that, in keeping with the OMB position, carcinogen risk assessments have been too "conservative," by the former head of EPA's Office of Health and Environmental Assessment (now an industry consultant), see E. Anderson in *Risk Analysis,* vol. 3, no. 4 (1983), p. 36.

pages 72–3: article by members of Syntex staff: See D. J. Paustenbach et al., in *Regulatory Toxicology and Pharmacology,* vol. 6 (1986), p. 284.

page 73: role of dioxin in biological process leading to appearance of tumor: This issue was discussed in some detail in my talk "Failure of the Environmental Effort," presented at the EPA Air and Radiation Program and Office of Toxic Substances seminar series, Jan. 12, 1988. See also Barry Commoner in *Environmental Law Reporter,* vol. 18, no. 6 (June 1988), p. 10195, and Barry Commoner in *Harper's* magazine, May 1988, p. 28.

page 75: EPA appointed task force of staff scientists: Although the original Nov. 1987 draft of the task force report has not been made public, I have received it as a member of the peer review group appointed to comment on the report. A modified version of the original draft (which reached the same conclusion as the original draft) has been published as *A Cancer Risk-Specific Dose Estimate for 2,3,7,8-TCDD* by the U.S. EPA Office of Health and Environmental Assessment, Washington, D.C., June 1988. The latter version was reviewed by the EPA Science Advisory Board (SAB) in 1988. Although the SAB findings have not yet been made public, the fact that the report's recommended reduction in the

1985 dioxin risk assessment has not been approved suggests that the SAB has rejected the task force's position.

CHAPTER 5. REDESIGNING THE TECHNOSPHERE

page 80: "Minicars make miniprofits" quotation: See *St. Louis Globe-Democrat,* May 14, 1971.

pages 80–1: John Z. DeLorean on switch to larger cars: See J. P. Wright, *On a Clear Day You Can See General Motors* (Grosse Pointe, Mich.: Wright Enterprises, 1979), p. 52. DeLorean provides the following data on profits from big cars (p. 177): "When I was with GM, a $300 to $400 difference between the building costs of a Chevrolet Caprice and a Cadillac DeVille, a bigger car, was small compared to a $3,800 difference in the sticker price. The difference in profit to General Motors on the two cars is over $2,000."

page 81: nuclear power would be "so cheap you wouldn't need to meter it": This statement has often been attributed to Lewis Strauss, former head of the Atomic Energy Commission, apparently based on a speech before the National Association of Science Writers in New York on Sept. 16, 1954, but that has been disputed. Although the statement may be apocryphal, it does accurately reflect the attitude of public officials and the industry in the early days of nuclear power.

page 81: detergents more profitable than soap: See Barry Commoner, *The Closing Circle* (New York: Knopf, 1971), p. 259.

page 83: substitution of plastic for leather: Data from the *Annual Survey of Manufacturers,* published by the U.S. Department of Commerce, show that in 1971, for example, the leather industry produced $6.25 of value added per person-hour of labor; $3.64 of value added per dollar of fixed assets (capital); $62.04 of value added per million BTUs of energy used. The comparable figures for the chemical industry, which produces plastics, are: $27.75; $0.80; and $4.85.

page 83: a difficult choice for the people of Tacoma: For a detailed account of the controversy over the Tacoma copper smelter, see P. Dorman in *Review of Radical Political Economics,* vol. 16, no. 4 (1954), p. 151. The quotation is from an article by D. J. Chasan in *The Weekly* (Seattle), April 24, 1983.

page 84: PCB exposure in 1933: See J. W. Jones and H. S. Alden in *Archives of Dermatology and Syphilology,* vol. 33 (1936), p. 1022. This article describes the severe effects of exposure to PCB on the workers in the first U.S. plant that produced this substance. The plant began operating in 1929, and by 1933, 23 of the 24 workers in the plant suffered from chloracne, which is a result of very high exposure to PCB.

page 85: Akio Morita quotation: See his article in *New York Times,* Oct. 1, 1989.

pages 85–6: changes in agricultural productivity: The input/output data are from *Economic Report of the President* (Washington, D.C.: Government Printing Office, January 1988).

page 86: changes in farm income: See *Economic Report of the President.* The data on prices received and paid by farmers are from table B-99, p. 421; the income data are from table B-96, p. 418.

page 87: Richard Rhodes quotation: See his op-ed article in *New York Times,* Nov. 24, 1989.

pages 87–8: decline of U.S. nuclear power industry: See *New York Times,* Mar. 23, 1989. See also J. L. Campbell, *Collapse of an Industry* (Ithaca, N.Y.: Cornell University Press, 1988), for a comprehensive discussion of the reasons for the industry's decline.

page 88: cost of Shoreham nuclear power plant; New York State purchase: See *New York Times,* May 27, 1988. In 1969 LILCO estimated the cost at $261 million; in 1973, when construction began, the cost was estimated at $503 million and the opening for 1977; in 1977 the cost was estimated at $1.2 billion and completion in 1980; the plant was actually completed in 1985 at a cost of $5.3 billion. See also Campbell, *Collapse of an Industry,* pp. 4–5, for a general discussion of escalating construction costs. Also see *New York Times,* May 12, 1988, for a discussion of the issues involved in the purchase of the plant by New York State.

page 89: output of chemical industry: See *Chemical and Engineering News,* June 8, 1987, p. 27.

page 89: EPA Toxic Release Inventory: See reference for chapter 2, pages 248–49.

page 89: toxic chemicals enter environment in various ways: See *National Survey of Hazardous Waste Generation Treatment, Storage and Disposal Facilities Regulated Under RCRA in 1981,* U.S. EPA, Washington, D.C., 1984. About two-thirds of the waste was injected into deep wells; about one-third was placed in landfills or surface impondments.

pages 89–90: cost of destroying toxic chemicals: The cost, $100 per ton, is an estimate based on current figures provided by Waste Management Inc. Costs vary considerably depending on the amount of waste in any given shipment.

page 90: chemical industry's profit in 1986: See U.S. Department of Commerce, *Survey of Current Business,* vol. 69, no. 7 (July 1989), p. 86.

pages 91–2: rising cost of producing oil in the United States: This is based on an analysis by the National Petroleum Council in *U.S. Energy Outlook: Oil and Gas Availability,* 1973. The curve that describes rising costs with time between 1950 and 1969 accurately predicts current costs as well. This issue and other aspects of the energy problem are discussed in Barry Commoner, *The Poverty of Power* (New York: Knopf, 1976).

page 93: Chicago Federal Reserve Bank report: See D. A. Aschauer, *Is Public Expenditure Productive?* Federal Reserve Board of Chicago Working Paper, 1988. A brief account of this paper appears in *Chicago Federal Letter,* Sept. 1988.

page 94: business decline resulting from beach pollution: See *Newsday,* Jan. 15, 1989.

pages 94–5: "Are Utilities Obsolete?"; cover story: See *Business Week,* May 21, 1984, and a related article on the slow growth in the demand for electricity in *New York Times,* May 22, 1984.

page 96: decentralization of electric power system: See, for example, an article on "do-it-yourself" electricity (i.e., cogeneration) in *New York Times,* June 24, 1984.

page 97: CBNS study of organic and conventional farms: See W. Lockeretz et al., *Agronomy Journal,* vol. 72 (1980), p. 65. This appears to be the first such comparative study of organic and conventional farming, and has been followed by many others. According to Gerald Stanhill, an organic agriculture researcher in Israel, a search of the scientific literature uncovered 226 comparisons of organic and conventional agriculture reported between 1979 and 1989. He reports: "Yields produced without any farm chemicals averaged 88% of those produced with the recommended optimum levels of application . . . [a] surprising refutation of the commonly held view that heavy applications of farm chemicals are essential for a productive agriculture." See his letter to the editor, *New York Times,* Nov. 13, 1989.

pages 97–8: National Research Council study of organic farming: See National Research Council, *Alternative Agriculture* (Washington, D.C.: National Academy Press, 1989).

page 98: CBNS study on ethanol production from crops: See *Economic Evaluation and Conceptual Design of Optimal Agricultural Systems for Production of Food and Energy,* Report to U.S. Department of Energy, Center for Biology of Natural Systems, New York, Dec. 16, 1981. See also Barry Commoner, "A Reporter at Large: Ethanol," *New Yorker,* Oct. 10, 1983.

page 99: National Science Foundation study: See F. P. Grad et al., *The Automobile and the Regulation of Its Impact on the Environment* (Norman: University of Oklahoma Press, 1975), p. 309.

CHAPTER 6. PREVENTING THE TRASH CRISIS

page 105: increase in per capita trash production: According to Franklin Associates Ltd., *Characterization of Municipal Solid Waste in the United States, 1960 to 2000,* U.S. EPA, Washington, D.C., July 1986, in 1960 the amount discarded per capita amounted to 2.65 pounds per day and in 1986, 3.58 pounds per day.

pages 105–6: data on beer bottles and beer consumption: See Barry Commoner in *Chemistry in Britain,* vol. 8, no. 2 (Feb. 1972), p. 52, table 6.

page 107: tipping fees since 1984: See *Waste Age,* March 1989, p. 101.

page 108: Westinghouse's entry into trash incineration: See *Wall Street Journal,* Nov. 14, 1986. The article quotes an analyst of the industry, Nicholas Hermann, as saying, "Westinghouse's nuclear business is clearly at a crossroads. They don't have many places to turn." The article continues with the comment: "However, the company's entry into several new energy-related businesses, notably the burgeoning waste-to-energy field, should soften somewhat the nuclear business decline."

pages 108–9: David L. Sokol quotation: The quotation is from his article in *Waste Age,* June 1988, p. 81.

page 109: incinerators sold between 1983 and 1987: These data were compiled by Michael Frisch of the CBNS staff from the following sources:

Office of Technology Assessment, *Materials and Energy from Municipal Waste, Resource Recovery and Recycling from Municipal Solid Waste and Beverage Container Deposit Legislation* (Washington, D.C.: Government Printing Office, 1979).

Radian Corp., *Municipal Waste Combustion Study: Characterization of the Municipal Waste Combustion Industry,* U.S. EPA, Office of Solid Waste, Office of Air and Radiation, and Office of Research and Development, Washington, D.C., 1987.

"Resource Recovery Activities," *City Currents,* Oct. 1983, Oct. 1984, Oct. 1985, Oct. 1986, U.S. Conference of Mayors, Washington, D.C.

"Waste Age 1988 Refuse Incineration and Refuse-to-Energy Listings," *Waste Age,* Nov. 1988, pp. 190–210.

page 109: family of 210 compounds commonly called dioxins: The term "dioxin," as now commonly used in general publications, applies to two related classes of compounds: polychlorinated dibenzo-p-dioxins and polychlorinated dibenzofurans. Within each class the various isomers differ in the number and molecular positions of the chlorine atoms.

pages 109–10: dioxin tests of emissions from Hempstead incinerator: The results are reported in a memorandum from Michael J. Dellarco, Coordinator, Dioxin Monitoring Program, Special Pesticide Review Division, U.S. EPA, Washington, D.C., dated June 18, 1980.

page 110: dioxin as environmental pollutant: J. E. Huff and J. S. Wassom have reported on a search of the early scientific reports about dioxin in *Environmental Health Perspectives,* Sept. 1973, p. 283. Evidence of the toxic effects of dioxin in chemical workers was reported as early as 1955; it occurred as a contaminant in the production of various chlorinated organic compounds such as PCBs and in herbicides such as 2,4,5-T. The earliest report of dioxin in the environment appears to be its detection in 1973 in fish and shellfish obtained in the areas of Vietnam treated with defoliants by the U.S. Air Force.

page 111: Commissioner of Sanitation quotation: See N. Steisel and P. D. Casowitz, "Incinerate New York Garbage," *New York Times,* Aug. 20, 1983.

page 111: dioxin emissions from European incinerators: A detailed compilation of dioxin emission data from incinerators worldwide has been published by M. R. Beychok in *Atmospheric Environment,* vol. 21 (1987), p. 29.

page 111: dioxin emissions not correlated with furnace temperatures: See Barry Commoner et al. in *Waste Management and Research,* vol. 5 (1987), p. 327, table 2.

page 112: cancer risk assessment for Brooklyn Navy Yard incinerator: See Camp, Dresser, and McKee, *Preliminary Draft Environmental Impact Statement for the Proposed Resource Recovery Facility at the Brooklyn Navy Yard,* prepared for New York City Department of Sanitation, Nov. 1982.

page 113: study on dioxin exposure via ingestion: See R. D. Kimbrough et al., *Health Implications of 2,3,7,8-Tetrachlorodibenzodioxin (TCDD) Contamination of Residential Soil,* Center for Environmental Health, Centers for Disease Control, Rockville, Md., 1983.

page 113: CBNS conclusion that maximum lifetime cancer risk would be 29 per million: See Barry Commoner et al., *Environmental and Economic Analysis of Alternative Municipal Solid Waste Disposal Technologies. III. A Comparison of Different Estimates of the Risk Due to Chlorinated Dioxins and Dibenzofurans from Proposed New York City Incinerators (Including a Critique of the Hart Report),* Center for the Biology of Natural Systems, New York, Dec. 1, 1984, table 1.

page 113: new consultant's estimated maximum lifetime cancer risk: 5.9 per million: See Fred C. Hart Associates, Inc., *Assessment of Potential Public Health Impacts Associated with Predicted Emissions of Polychlorinated Dibenzodioxins and Polychlorinated Dibenzofurans from the Brooklyn Navy Yard Resource Recovery Facility,* prepared for New York City Department of Sanitation, Aug. 17, 1984.

page 114: dioxin risk assessment eliminated from environmental impact statements: There is no assessment of the health risk from dioxin emissions in either statement, See Camp, Dresser, and McKee, *Draft Environmental Impact Statement for the Proposed Resource Recovery Facility at the Brooklyn Navy Yard,* Sept. 1984; and Camp, Dresser, and McKee, *Final Environmental Impact Statement, Proposed Resource Recovery Facility at the Brooklyn Navy Yard,* June 1985; both prepared for New York City Department of Sanitation.

page 114: new maximum lifetime cancer risk is 0.78 per million: See *Health Risk Assessment for the Brooklyn Navy Yard Resource Recovery Facility,* Health Risk Associates, Berkeley, Calif., April 30, 1987.

page 114: dioxin will hardly penetrate more than a few millimeters of soil: According to the 1985 U.S. EPA dioxin health-risk assessment, "Several authors have shown that the vertical movement of 2,3,7,8-TCDD in soil is negligible." EPA evaluations now assume that dioxin deposited in the air will remain in the upper 1 centimeter of soil. This contrasts with the Health Risk Associates assumption that dioxin will penetrate 10 centimeters of soil.

page 115: correcting for errors and absurdities, new maximum lifetime cancer risk is 12 per million: This computation was developed by Tom Webster of the CBNS staff, using the assumptions in the Health Risk Associates analysis, but correcting for the tenfold error in the depth of dioxin penetration of the soil and an underestimation of the dioxin dose that would be received from different sources.

page 115: a kind of technological tennis match: The New York Public Interest Research Group has played an important role throughout the continuing debate over the Brooklyn Navy Yard incinerator. It was an officially designated intervenor in the protracted DEC hearings on the plant's construction permit, in opposition to the permit.

page 115: most of the results range upward from one per million to 20 or more: See for example, table 3 in Barry Commoner et al., *Waste Management and Research,* vol. 5 (1987), p. 327.

page 116: no statistically significant relation between dioxin emissions and furnace temperatures or combustion efficiency: See above citation, table 2.

page 117: six-carbon rings occur in lignin: See Barry Commoner et al., "The Origins and Methods of Controlling Polychlorinated Dibenzo-p-dioxin and Dibenzofuran Emissions from MSW Incinerators," paper presented before the 78th Annual Meeting of the Air Pollution Control Association, Detroit, Mich., June 1985, table 9.

page 117: hypothesis that dioxin is synthesized in the incinerator: This hypothesis was described in the CBNS report cited earlier, *Environmental and Economic Analysis of Alternative Municipal Solid Waste Disposal Technologies. II. The Origins of Chlorinated Dioxins and Dibenzofurans Emitted by Incinerators That Burn Unseparated Municipal Solid Waste, and an Assessment of Methods of Controlling Them,* Dec. 1, 1984.

page 117: test of Prince Edward Island incinerator: The results of this test are reported in *The National Incinerator Testing and Evaluation Program: Air Pollution Control Technology,* Environment Canada, Sept. 1986. The significance of these results is discussed in Barry Commoner et al., *Waste Management and Research,* vol. 5 (1987), p. 327.

page 118: Saugus residents exposed to toxic materials: See D. Wallace, *A Preliminary Risk Assessment of the RESCO Incinerator Ashpile at Saugus, Massachusetts,* Center for the Biology of Natural Systems, New York, Oct. 31, 1989.

page 119: fly ash heavily contaminated with toxic metals: R. A. Denison of the Environmental Defense Fund, Washington, D.C., has summarized these data in his paper "Risks of Municipal Solid Waste (MSW) Incineration: An Environmental Perspective," presented at the Annual Meeting of the American Association for the Advancement of Science, San Francisco, Dec. 17, 1988.

pages 119–20: Peter Montague's comments on Falls Township incinerator: See *Hazardous Waste News,* Aug. 1, 1989 (a very useful newsletter edited by Dr. Montague).

page 121: H. Lanier Hickman, Jr., and Calvin R. Brunner quotations: For the Hickman quotation, see *Waste Age,* March 1988, p. 187. For the Brunner quotation, see *Waste Age,* March 1988, pp. 65, 70.

page 121: "Forget Love Canal . . . Arrest the NIMBY patrols": See *Newsweek,* July 24, 1989.

pages 121–2: study done for California Waste Management Board, quotations: See Cerrel Associates, Inc., *Political Difficulties Facing Waste-to-Energy Plant Siting,* Report Prepared for the State of California Waste Management Board, Technical Information Series, Los Angeles, 1984.

page 123: investment firm promoting bond issues, quotation: The quotation is from Kidder, Peabody Equity Research, *A Status Report on Resource Recovery as of December 31, 1987,* New York, April 29, 1988.

page 123: evidence of public impact on incinerator industry, quotation: See *Wall Street Journal,* June 16, 1988.

pages 123–4: description of LANCERS project, quotations: For a lucid analysis of the trash crisis, including a detailed description of the public controversy over the LANCERS project, see *War on Waste,* by Louis Blumberg and Robert Gottlieb (Washington, D.C.: Island Press, 1989). The quotations about the project are from this book.

page 125: some 40 proposed incinerators were blocked: This figure is based on reports compiled by *Waste Not,* a publication of Work on Waste USA, Canton, N.Y. This is a useful newsletter on trash problems.

page 125: striking parallel between demise of nuclear power and incinerator industry's difficulties: See article in *Wall Street Journal,* June 16, 1988. According to the article: "Critics profess to feel an eerie sense of *déjà vu* in the trend toward burning. In the 1990s, they say, this could become for municipalities what the nuclear plant building binge was to electric utilities in the 1970s. It plunged many into an economic and environmental swamp in which a few are still mired, their huge cost overruns unrecoverable from customers, their shareholder dividends shrunken or ended."

page 127: less fuel needed to produce molten glass from crushed glass than from sand: See R. F. Stauffer in *Resource Recycling,* Jan.–Feb. 1989, p. 24.

pages 127–8: percentage of crushed glass used by glass plants in United States and Holland: See C. Pollock, *Mining Urban Wastes: The Potential for Recycling,* Worldwatch Institute, Washington, D.C., 1987, p. 26, for a comparison of recycling rates in various countries. See C. Miller in *Resource Recycling,* Dec. 1989, p. 23, for data on the use of glass cullet in U.S. manufacturing of glass containers.

page 128: Los Angeles Times printed on paper that contains about 80 percent recycled material: See *Garbage,* Sept.–Oct. 1989, p. 12. The article cites other relevant percentages of recycled paper: *New York Times,* 10–15%; *Washington Post,* 0%; *Wall Street Journal,* 1.4%.

page 129: American Association of Newspaper Publishers on legislation mandating use of recycled newsprint: See *New York Times,* Sept. 22, 1989. While the Association's statement admonishes, "Newsprint producers should use recycled fiber to the greatest extent possible in the production of a quality product," it also asserts, "Regulating newsprint is regulating newspapers, and is intolerable in a free society."

page 131: survey of seventeen curbside recycling programs: See Institute for Local Self-Reliance, *Beyond 25 Percent: Materials Recovery Comes of Age,* Washington, D.C., 1989, pp. 12–13.

page 131: recycling system designed for Town of East Hampton, L.I.: See Barry Commoner et al., *An Intensive Trash Separation and Recycling System for the Town of East Hampton,* Center for the Biology of Natural Systems, New York, Dec. 1, 1986.

page 132: practical test set up with 100 volunteer households: The results of this test and the design of a full-scale intensive recycling system for the Town are described in Barry Commoner et al., *Development and Pilot Test of an Intensive Municipal Solid Waste Recycling System for the Town of East Hampton,* Final Report to New York State Energy Research and Development

Authority, Center for the Biology of Natural Systems, New York, Dec. 1988.

page 134: U.S. loses topsoil at rate of about 3 billion tons annually: See L. R. Brown et al., *State of the World, 1988* (New York: Norton, 1988), p. 173.

page 134: 1989 survey of communities with recycling programs: See Institute for Local Self-Reliance, *Beyond 25 Percent,* pp. 12–13.

page 134: recycling rates in Seattle: See T. Watson in *Resource Recycling,* March 1989, p. 28.

page 135: CBNS recycling study in Buffalo, N.Y.: See D. Stern et al., *Buffalo Curbside Recycling Pilot Program,* Final Report to Common Council of the City of Buffalo, Center for the Biology of Natural Systems, New York, Oct. 1989.

page 138: Warren County, N.J., incinerator: The problems encountered by this incinerator are described in articles in *New York Times,* Jan. 25, 1989, and in *Sunday Star Ledger* (Newark), Jan. 8, 1989.

CHAPTER 7. POPULATION AND POVERTY

page 143: Garrett Hardin quotation: The quotation is from Hardin's widely read article "The Tragedy of the Commons," *Science,* vol. 162, p. 1243.

page 143: Paul Ehrlich quotation: The quotation is from Ehrlich's best-seller *The Population Bomb* (New York: Ballantine, 1968), pp. 66–67.

page 143: head of NOW sought to enlist environmentalists: See *New York Times,* July 23, 1989.

page 144: Russell W. Peterson quotation: See *Interaction* (a publication of the Global Tomorrow Coalition), vol. 5, no. 2 (June 1985), p. 1.

page 145: Garrett Hardin on feeding hungry countries: The quotation is from Hardin's article in *Newsday,* July 27, 1989.

page 146: Club of Rome report: See D. H. Meadows et al., *The Limits to Growth* (New York: Universe Books, 1972).

page 147: this image is misleading: I must take some responsibility for promoting this misleading image, having, in a sense, embedded it in the title of my book about ecology, *The Closing Circle.* On rereading the discussion of the basic principles of ecology in chap. 2 of that book, I find insufficient emphasis on the role of solar energy in the ecosystem and hence on its thermodynamic "openness."

page 148: estimate of solar energy falling on land surface: See H. Thirring, *Energy for Man* (Weston, Conn.: Greenwood Press, 1968), p. 269.

page 150: data on beer bottles, pesticides, phosphate, and nitrogen oxides: These data are tabulated in Barry Commoner, *Chemistry in Britain,* vol. 8, no. 2 (Feb. 1972), p. 52.

page 152: analysis of role of environmental factors in developing countries: See Barry Commoner, "Rapid Population Growth and Environmental Stress," in *Consequences of Rapid Population Growth in Developing Countries,* Proceedings of a United Nations Expert Group Meeting, New York, August 23–26, 1988 (ESA/P/WP.110, June 29, 1989 [draft]).

pages 152–3: Bruntland Report, quotation: See World Commission on Environment and Development, *Our Common Future* (New York: Oxford University Press, 1987), p. 5.

page 153: total world production of food: See F. M. Lappé and J. Collins, *World Hunger: Ten Myths,* 2nd ed. (San Francisco: Institute for Food and Development Policy, 1982). See also P. R. Crosson and N. J. Rosenberg in *Scientific American,* Sept. 1989, p. 128. In response to the question "Will our species be able to feed itself when this steady state [i.e., the predicted stabilization of the world population at 10 billion] is reached?" the authors of this useful article respond: "The short answer is probably yes. World food production could grow significantly more slowly than the current rate, and there would still be enough food for 10 billion mouths by the time they arrive."

page 154: study of nutritional levels in India: See R. Revelle and H. Frisch, *The World Population Problem,* A Report of the President's Science Advisory Committee, Washington, D.C., 1967.

page 156: Garrett Hardin's "lifeboat ethic" quotation: See *Bioscience,* vol. 24 (1974), p. 561.

page 157: demographers have delineated a complex network of interactions: These relationships are succinctly described in E. A. Wrigley, *Population and History* (New York: McGraw-Hill, 1969).

page 158: Swedish demographic data: See D. J. Bogue, *Principles of Demography* (New York: Wiley, 1969), p. 59.

pages 158–9: demographic transition in industrialized and developing countries: See Nathan Keyfitz in *Scientific American,* Sept. 1989. The figure on p. 120 is a very useful plot of birth rate and death rate as a function of time in developed and developing countries. It clearly depicts the demographic transition in both groups of countries.

page 160: effect of colonization in Indonesia: See Clifford Geertz's incisive book *Agricultural Involution* (Berkeley: University of California Press, 1968).

page 161: colonization resulted in 1 billion excess in world population: See Nathan Keyfitz's article in *Journal of Social Issues,* vol. 23 (1967), p. 62.

pages 161–3: remarkable book by Mahmood Mamdani, quotations: See Mahmood Mamdani, *The Myth of Population Control* (New York: Monthly Review Press, 1972). This book is remarkable not only because of its courageous review of a prominent project but because of its valuable insights into the problem of population control.

page 164: relation between GNP per capita and birth rate in different countries: The data on GNP per capita are from R. L. Sivard, *World Military and Social Expenditures 1987–1988,* 12th ed. (Washington, D.C.: World Priorities, 1987). The birth-rate data are from *World Population Prospects: Estimates and Projections as Assessed in 1984,* United Nations, New York, 1986.

page 165: projected birth and death rates: The projections are based on the trends plotted in the figure in Keyfitz's article in *Scientific American,* cited above.

page 165: projected world food production: See Crosson and Rosenberg in *Scientific American,* p. 128.

page 167: abhorrent political schemes put forward in guise of science: Some of these schemes, such as application of the principle of "triage" to developing countries (in which developing countries too enfeebled to save themselves from starvation are deliberately abandoned), are discussed in Commoner, *The Closing Circle,* chap. 11.

page 167: one of Garrett Hardin's earlier astonishing proposals: The quotation is from Hardin's article "The Immorality of Being Soft-hearted," *Stanford Alumni Almanac,* Jan. 1969.

CHAPTER 8. ENVIRONMENTAL ACTION

page 171: the goddess Gaia: Gaia, the goddess of the Earth in Greek mythology, turns up in various guises in certain discussions more or less related to ecology. These range from the "Gaia hypothesis"—the view, based on the apparent tendency of some geochemical processes to establish stable equilibria, that the Earth is actually a living organism—to the notion that a homage to Gaia is the proper religion for ecologically minded people.

pages 171–2: Kirkpatrick Sale quotations: The quotations are from Sale's article in the Oct. 12, 1985, issue of *The Nation;* the article is adapted from Sale's book *Dwellers in the Land: The Bioregional Vision* (San Francisco: Sierra Club Books, 1985).

page 172: a major politician declared: The statement was made by Jesse Unruh, Democratic Leader of the State of California Assembly, in *Newsweek,* Jan. 26, 1970, p. 31.

page 173: Richard Nixon on environment, quotations: See *Nixon and the Environment* (New York: Taurus Communications, 1972), p. vii.

page 174: Agenda published by leaders of environmental organizations, quotations: See *An Environmental Agenda for the Future by Leaders of America's Foremost Environmental Organizations* (Washington, D.C.: Agenda Press, 1966). The self-proclaimed "foremost organizations" are National Resources Defense Council, Environmental Policy Institute, National Wildlife Federation, Environmental Defense Fund, Izaak Walton League of America, Sierra Club, National Audubon Society, National Parks and Conservation Association, Wilderness Society, and Friends of the Earth.

pages 176–7: the Acid Rain Roundtable: See *Chemical and Engineering News,* Nov. 18, 1985, for an account of the views developed by this group.

page 177: Jay D. Hair quotation: The quotation is from excerpts of a talk given by Dr. Hair at the GLOBESCOPE National Assembly in Portland, Ore., April 1985, as published in *Chemecology,* Oct. 1985 (a publication of the Chemical Manufacturers Association).

pages 177–8: World Wildlife Fund brochure, quotations: The quotation is from a brochure, *Conservation and Business Sponsorship,* published by the World Wildlife Fund Marketing Department, Washington, D.C. (no date). The brochure opens with a message from William K. Reilly, then president of the World Wildlife Fund, commenting that "many corporations . . . are forming new partnerships with respected conservation organizations because they recognize the benefits such relationships bring to their businesses as well as the environment."

page 181: older Washington-based organizations: These organizations comprise the list of "foremost environmental organizations" that drew up *An Environmental Agenda for the Future,* cited earlier.

page 182: Philip Shabecoff's article: See *New York Times,* Jan. 15, 1988.

page 183: EPA "Pollution Prevention Policy Statement," quotations: The statement is published in *Federal Register,* vol. 54, no. 16 (Jan. 26, 1989), p. 3845.

page 184: Mr. Reilly's congressional testimony, quotation: The quotation is from the statement submitted by William K. Reilly as testimony before a hearing of the Subcommittee on Transportation and Hazardous Materials, Committee on Energy and Commerce, U.S. House of Representatives, May 25, 1989.

page 184: President Bush as "preventionist": See *Washington Post,* June 8, 1989.

page 185: BACT provision of Clean Air Act: BACT is defined in Sec. 169(3) of the Clean Air Act as "an emission limitation based on the maximum degree of reduction of each pollutant . . . through the application of production processes and available methods, systems and technologies, including fuel cleaning."

page 186: EPA Region 10 agreed that BACT was applicable: This position is expressed in a memorandum from Gary O'Neal, Director, Air and Toxics Division, U.S. EPA Region 10, to Ronald

L. McCallum, Chief Judicial Officer, U.S. EPA. The memorandum concludes with the statement: "For all of the reasons stated above, Region 10 recommends that the PSD [Prevention of Significant Deterioration] permit be revised to include a Nitrogen Oxide limitation and to include requirements for recycling and source separation."

pages 186–7: letter from Resource Recovery Association, quotation: The quotation is from a letter dated April 12, 1989, to William K. Reilly from Gerard Lovery Lederer, General Counsel, U.S. Conference of Mayors. The letter refers to the National Resource Recovery Association as an "affiliate" of the U.S. Conference of Mayors. A copy of the letter was kindly provided by Mr. David Bricklin of the law firm representing the opponents to the Spokane incinerator.

page 187: Mr. Reilly's decision to reject application of BACT to incinerators: See PSD Appeal No. 88-12 in the matter of Spokane Regional Waste-to-Energy, Applicant, Before the Administrator, U.S. Environmental Protection Agency, Washington, D.C. Order Denying Review.

page 188: President Bush's June 12 announcement, quotation: President Bush is quoted in *EDF Letter* (Environmental Defense Fund newsletter), vol. 20, no. 3 (Aug. 1989), p. 1.

page 189: collaboration the outcome of dozens of meetings: See article on the Bush Clean Air Act proposals in *EDF Letter,* cited above.

CHAPTER 9. WHAT CAN BE DONE

pages 191–2: view of Earth First!: The statement is quoted from an Earth First! flyer describing its aims.

page 196: such a design has been described: See P. B. Weisz and J. F. Marshall in *Science,* vol. 206, p. 24. This paper, by members of the staff of the Mobil Research and Development Corporation, concludes that the production of ethanol from corn consumes more energy than the resultant ethanol contains. This result is accomplished, in part, by assuming that all the energy used to produce a bushel of corn represents energy required for ethanol production. In fact, however, about 40% of the corn remains, after ethanol production, as a useful feed, so that the energy used to produce it should not be charged to ethanol production. Moreover, the Mobil scheme involves the use of an energetically inefficient ethanol distillery. When these faults are corrected, it is evident that ethanol production represents a net gain of energy.

page 198: natural gas as a bridging fuel: The important role that natural gas can play in bridging the transition from the present fossil-fuel energy system to one based on solar energy is discussed in detail in Barry Commoner, *The Politics of Energy* (New York: Knopf, 1979).

page 200: computer models of global warming: The most frequently cited model is that described by J. Hansen et al. in *Journal of Geophysical Research,* vol. 93, no. D8 (1988), p. 9341. The accuracy of such models depends, of course, on the validity of the assumptions on which they are based. A major assumption in global warming models is that airborne water vapor would amplify the warming effect of carbon dioxide and other gases. Critics have challenged this assumption and hence the model's predictions. A recent article by V. Ramanathan et al. in *Nature,* Dec. 14, 1989, reports actual measurements of the effect of airborne water vapor on atmospheric temperature. The results confirm the assumed amplification effect.

page 200: effort to replace CFCs: For a readable account of the ozone problem and remedial actions, see A. Makhijani et al., *Saving Our Skins,* Environmental Policy Institute and Institute for Energy and Environmental Research, Washington, D.C., Sept. 1988.

page 202: Worldwatch Institute estimate: See L. R. Brown et al., *State of the World, 1988* (New York: Norton, 1988), p. 183.

page 202: Dr. Ullman's estimate: The problem of restoring U.S. railroads and the cost of doing so are discussed by John E. Ullmann in *The Prospects of American Industrial Recovery* (Weston, Conn.: Greenwood Press, Quorum Books, 1985), p. 183. I am indebted to Dr. Ullmann for a current estimate that it would cost about $1.5 million per mile to modernize 66,000 miles of railroad, or about a total of $100 billion.

page 203: fixed capital in petrochemical industry: See *U.S. In-*

dustrial Outlook 1989, U.S. Department of Commerce, Washington, D.C., p. 84. In 1987 the values of current-cost gross stock of fixed capital in the two industrial categories that approximately encompass the petrochemical industry, chemicals and allied products and rubber and plastic products, were $193.0 billion and $48.3 billion, respectively. Since some inorganic products of the chemical industry may be used outside the petrochemical industry, the total is rounded out to $200 billion.

page 203: capital invested in coal mining, oil and natural gas production, and petroleum refining: See above citation, p. 84. In 1987 the values of current-cost gross stock of fixed capital in these categories were: coal mining, $47.3 billion; oil and natural gas production, $315.2 billion; petroleum and coal products, $102.4 billion. Allowing for some growth in assets since 1987, the sum is rounded to $500 billion.

page 203: subsidies to U.S. farmers: See *Statistical Abstract of the United States, 1989,* U.S. Department of Commerce, Washington, D.C., p. 635.

page 204: Section 102C of National Environmental Policy Act: The relevant parts of this section state that, with respect to governmental actions affecting the environment, statements must be prepared regarding (in addition to environmental impact):

"(iii) alternatives to the proposed action;

"(iv) the relationship between local short-term uses of man's environment and the maintenance and enhancement of long-term productivity."

pages 205–6: development of integrated computer chips: See *DOD Photovoltaic Energy Systems Market Inventory and Analysis,* Federal Energy Administration, Washington, D.C., 1977, vol. 1, pp. ix, 10–11. In 1962, about 160,000 chips were produced for DOD at a cost of about $50 each. By 1968, annual production had increased to about 120 million units, each costing about $2.50.

page 206: cost of producing photovoltaic cells in 1970s: See Commoner, *Politics of Energy,* p. 35. In the early 1970s, photovoltaic cells were in small-scale commercial production, chiefly for use on satellites. In 1976, they were available on the open market at a price of about $15 per peak watt. The current price is about $4–$5 per peak watt.

page 207: stratified charge engine: See F. P. Grad et al., *The Automobile and Regulation of Its Impact on the Environment* (Norman: University of Oklahoma Press, 1975), pp. 299ff.

page 207: federal government purchase of cars and trucks: In 1986 federal purchases of automobiles amounted to $1.6 billion; purchases of trucks amounted to $5.7 billion. See U.S. Department of Commerce, *Survey of Current Business,* vol. 69, no. 11 (Nov. 1989), pp. 7–8.

page 208: Alar produces cancer in test animals: See B. H. Sewell et al., *Intolerable Risk: Pesticides in Our Children's Food* (Washington, D.C.: National Resources Defense Council, 1989), Summary, p. 2.

pages 208–9: but Alar broke out of this pattern: See *New York Times,* May 16, 1989, for an account of the decision to take Alar off the market.

page 209: Jeffrey Hollender's do's and don'ts: See *Newsday,* Aug. 10, 1989.

CHAPTER 10. MAKING PEACE WITH THE PLANET

page 213: Mr. Reilly's article, quotation: See William K. Reilly, "The Greening of EPA," *EPA Journal,* vol. 15, no. 4 (July–Aug. 1989).

page 214: USX plan to build modern steel mill: The company's movement into the oil business is discussed in *Business Week,* Oct. 6, 1986, p. 26. The shelving of the plan to build the Conneaut plant is discussed in an article on the company's retreat from steelmaking in *New York Times,* Dec. 2, 1979.

page 215: USX's business in oil and steel; Robert Reno quotation: See *Newsday,* March 14, 1986. See also *Business Week,* Oct. 6, 1986, regarding relative investment in oil and steel.

page 215: John Swearingen quotation: See *Chicago Tribune,* April 19, 1978.

page 216: Berle and Means quotations: See Adolph A. Berle and Gardiner Means, *The Modern Corporation and Private Property* (New York: Macmillan, 1932). The first quotation is from p. 46 and the second from p. 356.

page 217: Catholic bishops' pastoral letter, quotation: The quotation is from par. 110 of the letter, "Economic Justice for All: Catholic Social Teaching and the U.S. Economy," as published in *Origins,* vol. 16, no. 3 (June 5, 1986).

page 217: Pope John Paul II quotation: The quotation is from excerpts from the English version of the Pope's encyclical "On Human Work," published in the *New York Times,* Sept. 16, 1981. This statement is cited in par. 113 of the bishops' letter in support of its general position.

page 218: bishops' letter quickly disappeared from public sight: This statement is based on the results of a computer search of the *New York Times* from the date of publication of the letter (June 3, 1986) through Oct. 1989.

page 219: Lord Beeching quotation: This statement is quoted from Barry Commoner, "The Promise and Perils of Petrochemicals," *New York Times Magazine,* Sept. 25, 1977.

page 222: major projects abandoned in Soviet Union: According to the Japanese news service, Kyodo, A. Lapshin, the Soviet deputy minister of atomic energy, stated in an interview that 20 nuclear power plants have been shut down or their construction halted since the Chernobyl accident. (I am indebted to Eric Johnson for this information.) That this change in official policy, which has strongly favored nuclear power, is at least in part a response to environmental pressure is evident from an article by a group of members of the Ukrainian Academy of Science highly critical of nuclear power plans. This is reported in *Science,* vol. 239, p. 1090, which also notes, from an article in *Pravda,* that a nuclear power plant at Krasnodar was canceled after the public raised questions about the adequacy of its protection against earthquakes.

page 223: Manfred Bienefeld quotation: See his article in *Monthly Review,* vol. 41, no. 3 (July–Aug. 1989), p. 9.

page 224: there is glee and smugness: The major example of this approach is Francis Fukuyama's article "The End of History?" *National Interest,* Summer 1989. The article speaks of the end of history, which is supposedly occasioned by "the total exhaustion of viable systematic alternatives to Western liberalism," and will be "a very sad time" characterized by "the endless solving of technical problems, environmental concerns and the satisfaction of sophisticated consumer demands." This is hardly an accurate description of the situation in the United States, where, as we now know, the effort to solve environmental problems has failed, and where consumer demands (largely unmet) are no more sophisticated than decent housing and medical care. And it is, of course, ludicrous to declare the end of history in 1989—the year when the changes in Eastern European history have broken all records for scope and speed.

page 225: the Valdez Principles: The quotation is from the opening paragraph of the Principles, as stated in a press release from the Coalition for Environmentally Responsible Economics, Sept. 7, 1989.

page 230: Green Party in 1983 election: See H. Mewes in *Environment,* vol. 27, no. 5 (June 1985), p. 12.

page 231: resolution condemning slaughter of frogs: For an account of this event see Diane Johnstone's article in *In These Times* (Chicago), July 24, 1985.

page 231: Green Party split: The Green Party decision to join a coalition government in Hesse is discussed in *New York Times,* Oct. 28, 1985.

page 232: helping to reverse Italian Communist Party's support for nuclear power: For a number of years the party gave its tacit support to the government's plan to build as many as eight new nuclear power plants in Italy, beginning at Montalto de Castro. The issue was, however, constantly debated within the party, culminating in a vote at the February 1986 party congress that narrowly (by 17 votes out of some 2,000) failed to adopt an antinuclear position. The closeness of that vote, and the Chernobyl accident soon afterward, led the party to adopt a position against even the completion of the plant at Montalto de Castro a few months later.

page 237: some $30 billion flows from poor countries to rich ones: See World Commission on Environment and Development, *Our Common Future* (Bruntland Report; New York: Oxford University Press, 1987), p. 69. Also see L. R. Brown et al., *State of the World, 1988* (New York: Norton, 1988), p. 184.

page 238: worldwide military budget: See R. L. Sivard, *World*

Military and Social Expenditures, 1987–1988, 12th ed. (Washington, D.C.: World Priorities, 1987), p. 7.

page 238: U.S. portion of world GNP: See above citation, p. 46, for the relevant data.

page 240: rapid growth of military research and development: See M. Renner in Brown, *State of the World, 1988,* p. 32.

page 240: relation between military budget and economic productivity: Data for United States, Canada, United Kingdom, France, West Germany, Italy, and Japan on the average annual rate of increase in labor productivity for the period 1971–1985 are provided by D. A. Aschauer, *Is Public Expenditure Productive?* Federal Reserve Board of Chicago Working Paper, 1988. Data on the military budget expressed as percent of GNP for 1984 are provided by Sivard, *World Military and Social Expenditures.*

pages 240–1: impact of military budget on Soviet economy: In January 1989, President Gorbachev reported to a group of visitors that the Soviet military budget would be reduced by 14 percent. See *New York Times,* Jan. 19, 1989.

INDEX

ACKNOWLEDGMENTS

MUCH OF THE material discussed in this book derives from my activities in the Center for the Biology of Natural Systems over the last twenty-three years. I am indebted to the present CBNS staff for helping to develop much of the information presented in the book: Mark Cohen, Holger Eisl, Michael Frisch, James Quigley, Anita Sadun, Alex Stege, Deborah Wallace, and Tom Webster. I am especially grateful to Sharon Clark Peyser, CBNS's Administrative Coordinator, for her invaluable help in enabling me to complete this project—and many earlier ones as well. And I wish to thank the staff of Pantheon Books for effectively expediting the publication process. Finally, most of the ideas developed in this book, and in particular the relationships among them, are the outcome of trenchant comments and incisive critiques generously contributed by my wife, Lisa Feiner.